Praise for *Waves of Thoughts*

"*Waves of Thoughts* is a deeply researched and beautifully written journey through humanity's technological rise. It combines curiosity, intellect, and poetic storytelling to show how breakthroughs ultimately uplift society. Balaji emerges not just as an author, but as an inspiring and thoughtful human being."

Frank Wolf
Former Vice President, Alstom, Central Europe

"It is perfect for anyone with a curious mind, enticing readers to explore a finely curated set of examples that illustrate how ideas spark other ideas and how progress is rarely linear. Through compelling examples, it reveals how many world-changing innovations grew from humble beginnings—born of necessity, shaped by ingenuity, and driven by human creativity. An illuminating and enjoyable read."

Todd Connolly
Vice President, Global Brand Transformation

"*Waves of Thoughts* is an engaging journey through the evolution of human ideas, and it's clear from the very first chapters just how thoroughly researched this book is. Each topic—from ancient innovations to modern technologies—is presented with depth, clarity, and a genuine appreciation for the thinkers who shaped our world. What stood out to me most was how seamlessly B R Jallipalli blends history, science, and storytelling. The book never feels heavy, yet you can tell it's built on an enormous amount of careful study and reference work. It's a fascinating read for anyone who loves understanding not just what humanity created, but how those ideas came to life."

Stuart Patch
Fractional CFO & Financial Educator, Patchway Services; Formerly Global Finance Director, Cummins Generator Technologies

"A compelling and thought-provoking outline that explores how simple inspirations evolve into transformative ideas, while also inviting reflection on how emerging technologies like AI may shape human creativity and learning."

Colin Shaw

Former EGM of People and Strategy, and CEO of Training Academy

"This book offers a unique and deeply personal perspective, mixing vision and passion; it is rooted not in the technical wonders of the world but rather in the thoughts, needs, and aspirations that led to those innovations. It can be educational and is always accessible. Overall, an almost contemplative and uplifting line of thought, carried by a gentle, rested mind not ever afraid of what the future holds."

Arnaud Charpentier

CEO, Publishing Technology & Solutions (PTS);
Former Director, Finance Business Operations, Tesla

"*Waves of Thoughts* explores the remarkable journey of humanity by examining the pivotal ideas that have driven our most significant cultural, scientific, and technological advancements."

Dr Mahesh V Gantasala

Specialist Prosthodontist, Sydney, Australia

"A unique read of lateral thinking showing the linkage of people's ideas to the progress in our world. The breadth of knowledge and understanding of subject matter is impressive."

Dr Louis Shulman

Radiologist, Sydney, Australia

"A comprehensive journey through how thoughts, creativity, and relentless passion shaped the world we live in today."

Cheryl Khoo

HR Manager, Vellocet Clean Energy;
Former Global HR Generalist (Asia Pacific)

Waves of Thoughts

Thoughts That Changed the World

B R Jallipalli

Waves of Thoughts: Thoughts That Changed the World

A catalogue record for the book is available from the National Library of Australia.

ISBN: 978-1-7642561-4-8
Published by Thought Waves Press
Sydney, Australia

First edition, 2026
Typeset in Garamond 12 pt by Istvan Szabo, award-winning formatter and graphic designer

Printed in the country of purchase by Amazon KDP. Disclaimer:
This book is based on historical records, published works, and general knowledge. While every effort has been made to ensure accuracy, the author and publisher assume no responsibility for errors or omissions, or for the interpretation of information contained herein.

Waves of Thoughts

Thoughts That Changed the World

"One idea can change the course of humanity"

— B R Jallipalli —

Thought Waves Press

Dedication

To the thinkers, inventors, and visionaries whose ideas reshaped our understanding of the world, and whose courage to imagine has guided humanity forward.

To the quiet minds who dared to ask: *What if?*

To the restless thinkers who challenged the known, and to the builders who transformed the unseen into reality. You did not always make headlines, but you made history. You pursued questions others ignored, recognized connections others missed, and pressed on when answers seemed distant.

Waves of Thoughts was born from deep reverence for the currents of human thought that carried us from fire to flight, from whispers to wireless, from wheels to quantum wonder.

This book honors not only the ideas that shaped civilization, but the minds bold enough to conceive them.

May your spirit of wonder continue to inspire generations to come. May these pages echo your legacy and awaken the next great thought, waiting quietly in someone, somewhere.

With gratitude,

B R Jallipalli

Table of Contents

PART 2

PART 5

Duvvuri Subbarao served as Governor of the Reserve Bank of India from 2008 to 2013. Over a distinguished career spanning public administration, international development, and economic policy, he has held senior roles in the Government of India and at the World Bank. He is currently a Fellow at Yale University. He studied at IIT Kanpur, Ohio State University, and MIT, where he was a Humphrey Fellow, and is also the author of *Who Moved My Interest Rate* and *Just a Mercenary*.

FOREWORD

Throughout history, the trajectory of human progress has been shaped not solely by circumstance or necessity, but by the transformative power of ideas. Time and again, a single act of thought, often quiet, tentative, and even unconscious, has altered the course of societies and redefined the boundaries of what is possible.

In *Waves of Thoughts* – Thoughts That Changed the World, Balaji Jallipalli undertakes an ambitious and thoughtful exploration of this enduring truth. Spanning vast stretches of human history, the book traces the evolution of ideas that began as moments of curiosity and grew into innovations that reshaped civilization. From humanity's earliest mastery of fire to the frontiers of quantum computing and artificial intelligence, the narrative reminds us that every technological advance originates in wonder, a question posed, an assumption challenged, an imaginative leap taken.

This book is distinctive not just because of the broad scope of the story it tells but also because of the coherence with which that story is told. Rather than presenting inventions as isolated milestones, the author frames them as interconnected currents of thought, each insight influencing the next, each breakthrough sending ripples across generations. This approach invites the reader to engage not just with outcomes, but with the intellectual context that made them possible.

At a time when progress is often measured solely by results, this book compellingly redirects our attention to beginnings: to the formative moments when ideas emerge and take shape. In these moments, the defining qualities of humanity—curiosity, resilience and the courage to question—are revealed with particular clarity.

Waves of Thoughts is therefore more than a historical account of invention. It is a meditation on the human capacity to imagine, to question, and to shape the future through thought. It affirms that progress is rarely linear, but it is rarely accidental.

Why should one read this book? Because it offers something increasingly rare: perspective. In a world saturated with information but short on understanding, *Waves of Thoughts* restores context. It allows us to see human progress not as a series of self-evident achievements, but as outcomes of long and patient intellectual pursuit. In students, it cultivates intellectual humility. In professionals and leaders, it sharpens strategic judgment by revealing how systems actually evolve. In the general reader, it sparks wonder—not at novelty alone, but at the sustained human effort behind it.

Above all, *Waves of Thoughts* reaffirms the enduring power of ideas to illuminate the past and inform the way forward.

Hyderabad, India
January 2026

Duvvuri Subbarao
Governor (2008–13)
Reserve Bank of India

INTRODUCTION

WAVES OF THOUGHTS

Thoughts That Changed the World

I close my eyes, and when I open them again, the world has changed.

The noise of engines, the glow of streetlights, and the restless pulse of modern life recede. The year is 1825. The ground shifts unevenly beneath hooves and wooden wheels as a horse-drawn carriage rattles past. There are no steel tracks, no roaring locomotives, no engines conquering distance. To travel far is to walk, to ride, or to trust uncertain winds. A thousand miles is not a journey, but a commitment measured in months.

At the docks, wooden ships rest against the tide. No engines churn; no smoke stains the sky. These fragile vessels carry both trade and hope across unforgiving waters. Above me, the heavens remain empty. Flight is still a myth, not a machine. Were I to claim that humanity would one day cross oceans in hours, the world before me would laugh. Yet I would know the truth. I have seen where we are headed.

I open my eyes once more. The present returns.

Airports thunder. Trains carve through continents. Satellites blink across silent skies. Humanity has compressed distance into memory. What has changed is not technology alone, but the thoughts that shaped it. Ideas, once quiet and private, became engines, equations, and revolutions.

Progress does not arrive overnight. Every invention begins as a question that refuses to fade.

A World Built from Sparks

We live surrounded by tools so familiar that their origins have become almost invisible: a pen, a light bulb, a road, a wire. Each began not as a physical object, but as a thought. Fragile. Private. Unproven. The world we now inhabit is a gallery of such thoughts, transformed into reality through persistence.

Thomas Edison, often celebrated as an emblem of invention, once observed:

"I never had an idea in my life. My so-called inventions already existed in the environment. I took them out."

His words reveal a subtle truth. Innovation is less an act of divine inspiration and more a disciplined capacity to perceive patterns that others overlook. Thought becomes invention when curiosity is sustained by endurance.

This book is not solely about machines. It is about the evolution of thought itself, and the way ideas travel through generations, shaping societies and redefining what humanity believes to be possible.

How Thought Moves Through Time

Thought behaves like a wave. It rises, collides, dissipates, and returns in altered forms. From the first spark struck from stone to the coded languages that now guide intelligent machines, progress has never followed a straight line. Ideas recur across eras, resurfacing with renewed relevance.

This journey does not adhere to a rigid timeline. Instead, it moves through history as thought itself does: sometimes ancient, sometimes modern, always interconnected.

The purpose is not to overwhelm with technical detail, but to offer a path that invites reflection, whether for a scholar or a curious reader. Time and again, the simplest truths have produced the most profound transformations.

A single thought resembles a seed. Unremarkable in appearance, yet capable of growing roots, branches, and entire forests. Civilization itself is constructed from ideas that once existed quietly within a single mind.

Every revolution begins with someone willing to think differently.

Our Place in the Continuum

As we explore these waves of human thought, we encounter individuals who bridged imagination and reality, perceiving possibility where others saw none. Their contributions form enduring threads in a vast tapestry that continues to unfold.

Thought is never static. It travels across generations, linking ancient hands that carved symbols into stone with modern minds designing digital landscapes. It is the invisible current that propels humanity forward.

We are not merely beneficiaries of past ideas. We are active participants in the next wave.

This book traces a continuum, from the earliest questions that illuminated the ancient world to the algorithms shaping our present age. With curiosity as our compass, we follow the streams of language, perception, invention, healing, and creation to understand how thought ultimately becomes the world.

A Focused Journey

The ocean of human thought is too expansive for any single volume. This work does not attempt to chart every current, but instead follows the most consequential ones: expressions that connected societies, tools that extended human capability, and visions that redefined time and space.

From carved stone to quantum circuits, ideas do not vanish. They accumulate, refine, and build upon one another.

Let us explore the waves that carried us to this moment, and the ideas still forming beneath the surface.

Welcome to the journey.

FOUNDATION CHAPTER

The Roots of Wonder

Where Curiosity Became Human

A spark. That is how it begins. Not with tools. Not with fire. It begins with a question, a glance at the stars, a restless feeling at the edge of a cave. Long before steam hissed and silicon pulsed, there was the first quiet Why? Fragile, private, persistent. A small shift from reacting to imagining. Curiosity is not an accessory to intelligence. It is the force that animates thought. From the spinning wheel to the worldwide web, progress has rarely arrived through brute force. It arrives when people decide to read patterns, to test them, and to shape what they see. This chapter does not tell the story of machines. It outlines the story of thought itself, the first motion from instinct to inquiry, the dawn of asking that set everything else in motion.

How Our Ancestors Began Thinking Differently

Two million years ago, a hominid chipped stone into a tool. The action mattered not only because it solved a problem, but because it implied foresight. To fashion a flake is to imagine a future use. That is the quiet birth of planning. By roughly 300,000 years ago, Homo sapiens displayed signs of a richer inner life. Marks in ochre at Blombos Cave and engraved patterns on shell and bone are not idle decorations. They are evidence of minds making symbols, of people testing ways to hold meaning outside themselves. Around shared fires, heat became more than warmth. Stories gathered there. Faces lit by flame listened and learned. Memory moved from the individual into the group.

These were not merely survivors. They were architects of thought.

From Symbols to Shared Memory

Between painted caves and carved tusks, human thought stepped beyond description and into meaning. The images at Lascaux and Sulawesi are more than records of hunts.

They are experiments in representation, perhaps rituals, perhaps teaching tools. The point is that someone chose to make an image that could carry a message across time. Language likely grew in those same moments. It began as a warning and coordination, but it became so much more. Between 100,000 and 50,000 years ago, groups refined ways to transfer knowledge.

A hunter could teach a child techniques she had never seen. Ideas began to survive past a single lifetime. Burial practices from thirty thousand years ago, with beads and offerings, suggest people were thinking about identity and memory, not simply bodies. These rituals are the first signs of reflection on existence itself. Curiosity had become culture.

Mapping Thought into Space

To wonder is one thing. To map wonder is another. Early engravings that resemble star charts, and etched tusks from twenty-five thousand years ago that look like maps, show a new capacity: translating the world into patterns. Spatial thinking turned landscape into plan, path, and possibility. Stone alignments and city grids are later expressions of the same impulse that first drew lines in dust. Mapping gave humans a tool to predict, to organize, and to imagine worlds beyond the immediate horizon. It was a new way of making thought portable.

The Moment We Began to Ask Why

Some questions change everything. When humans began asking Why, they turned curiosity inward. Agriculture, beginning about ten thousand years ago, required prediction. People asked why seasons varied and why crops failed. Myth and measurement grew side by side. Early civilizations in Sumer and elsewhere began to model the world through story and calculation, gods and records. Technological leaps followed the urge to understand. The wheel solved a question about motion. Metallurgy answered questions about transforming substances. These inventions were not accidents. They were replies to persistent questions. A single hand -shaping stone became a mind -shaping possibility. Across continents and centuries, humans tested, mapped, argued, and imagined. From ochre on cave walls to code in circuits, the same current runs through history: curiosity.

FROM ZERO TO NOW

A Snapshot of Human Progress

Sometimes we forget how recently the world we take for granted was built. A few lifetimes ago, almost everything around us simply did not exist.

600 years ago
No books in every home; printed knowledge is largely inaccessible to ordinary people

250 years ago
No steam engines are reshaping transport and industry.

190 years ago
No practical photography to capture everyday life

150 years ago
No telephones to carry voices instantly across cities and continents

140 years ago
No electric light to extend the day or reshape human habits

130 years ago
No cars enabling personal and rapid travel.

125 years ago
No radio to broadcast ideas across nations

120 years ago
No human flight to compress oceans into hours

Note: Time spans are rounded to the nearest decade for clarity. For exact years and an extended timeline of innovations, refer to Annexure 1 at the back of the book.

A Guide to What Follows

This book is arranged like a tide, not a timeline. You will find five parts. Each part is a current of thought that builds on what came before and points toward what follows. Think of each chapter as one wave within a larger sea. Read straight through, or dip in where curiosity pulls you. Either way, you will meet examples, human moments, and the small practical moves that grew into systems.

Part 1: Dawn of Thought: The Foundations of Human Progress
Here we trace the first moves from survival to symbol. Expect to encounter the quiet inventions that altered perception: language, storytelling, art, music, and early technologies that allowed thought to be shared and stored. This part shows how a private spark became a communal flame.

Part 2: Machines of the Mind: Inventions That Transformed Society

In this section, ideas become mechanisms. Wheels, engines, ships, rails, and flight recorded distance and time. Engineers borrowed from nature, and mechanical systems began to shape economies and everyday life. Read how design and imitation amplified human reach.

Part 3: Digital Tsunami: From Computers to Consciousness

This part follows computation and networks. Code became a new language of thought, and connectivity reshaped culture and choice. Expect stories of visionaries and rapid shifts that moved thinking from human brains to distributed systems.

Part 4: Health, Humanity, and the Science of Life

Here we explore the slow, careful work that made the invisible visible. From microscopes to imaging, from vaccines to precision surgery, this section shows how understanding the body transformed medicine and our expectations about life and repair.

Part 5 Thoughts That Built Civilization: Economy, Shelter, and Convenience

Small practical ideas about trade, money, food, and shelter became the scaffolding of cities and daily life. This final part shows how convenience and systems shape culture, consumption, and the patterns that define modern living.

How to Read This Book

Treat chapters as invitations. Each is short enough to hold attention and long enough to carry a thought from spark to consequence. I will point to how observation became an experiment and how experiment became a system. The aim is not an exhaustive history. It is a clear thread: thought leads to trial, trial leads to invention, and invention reshapes how we live and think.

Before We Begin

This book asks you to notice the ordinary as the origin of the extraordinary. A pen is not merely an object. It is the endpoint of a particular thought. Civilization is a forest of small seeds. This foundation chapter is your map. The chapters that follow are the stories of those seeds. Welcome to the journey.

PART 1

Dawn of Thought

The Foundations of Human Progress

Chapter 1
When a Spark Becomes a Symphony

The Journey of a Thought

Thought rarely announces itself with spectacle. It begins quietly, as a faint signal in a network of neurons that have spent a lifetime storing memories, forming associations, and predicting outcomes. A single electrical impulse may seem insignificant, yet history shows that the smallest cognitive spark can reshape societies, invent languages, and launch revolutions.

We often admire inventions like airplanes or political movements that shift nations, but we rarely pause to ask a simpler question: what was the first internal moment that led to them? Before sketches, speeches, and blueprints, there was an internal event in someone's mind. A thought. To understand human progress, we must begin there.

Where Thoughts Come From

The brain contains roughly eighty-six billion neurons. Each connects to thousands of others through synapses that strengthen or weaken over time. Thoughts are not random flashes; they emerge when patterns of memory, emotion, and sensory input align in new ways. You might be sitting still, hearing a faint sound in the background, and suddenly a memory appears. The hippocampus retrieves stored experiences. The association cortex gives shape to the fragments. A coherent idea forms.

Most thoughts never rise to conscious awareness. They drift beneath the surface, like background waves that never reach shore. But occasionally, one breaks through and demands attention. Awareness is not simply noticing a thought; it is evaluating it. The brain weighs relevance, emotional value, and potential outcomes. A thought that excites, worries, or inspires is more likely to progress into action.

When Thought Turns Into Action

There is a moment when an internal impression becomes external movement. The shift is subtle. Muscles activate, attention sharpens, the body prepares. A person reaches for a pen. A voice speaks a sentence out loud. A hand sketches an idea for the first time. The invisible becomes visible.

Some ideas arise quickly, formed in seconds. Others take years of accumulation. Newton did not discover gravity because of a falling apple alone; the story is often romanticized. He had been thinking about motion and celestial mechanics long before the observation aligned with a question he was already trying to answer. The same can be said of Gandhi or Steve Jobs. Ideas that change history are rarely spontaneous bursts. They are often long processes that eventually cross a threshold.

Ideas That Spread

A private thought becomes influential when it travels beyond the individual mind. Writing, speech, and collective memory turn personal insight into shared action. Ideas spread through education, imitation, storytelling, or conflict. They gather force the way rivers gather speed, joining other waters.

When enough people adopt a thought, it becomes culture. When culture persists across generations, it becomes history. The transformation from a single thought to a movement is neither automatic nor guaranteed; most ideas fade. The ones that remain do so because they find structure, relevance, and carriers willing to preserve them.

The Continuous Cycle

Action is not the end of a thought's journey. Results create new perceptions, which create new thoughts. The cycle continues: perception influences thought, thought shapes emotion, emotion drives decisions, decisions produce actions, actions reshape perception. This loop forms the basis of human adaptation and progress.

Civilization is not built on ideas alone, but on repeated cycles of thought becoming action and then thought again. Every tool, belief system, medical breakthrough, or artistic creation began as a silent spark before becoming part of the external world.

A Quiet Beginning

You are not simply reading about inventions or historical achievements. You are observing the internal mechanism that precedes all human change. A thought is not dramatic when it begins. It is quiet, private, and sometimes fragile. Yet inside it exists the potential to rearrange reality.

Every chapter that follows traces how those sparks shaped humanity.

This is where the journey begins.

Chapter 2
The First Sunrise: Language, the Bridge of Thought

Scientists believe language emerged gradually, not with sudden invention but through countless generations discovering that sound could carry intention. Long before cities or agriculture, early humans learned that survival depended on connection. A warning cry meant danger. A calming murmur reassured a child. Over time, these sounds evolved from instinct to expression.

Thought alone was not enough. Ideas confined within a single brain remain invisible. Without a way to share them, there can be no memory beyond one lifetime. So language became the first true bridge between minds, a human-made channel for transmitting knowledge, emotion, and imagination.

From Instinct to Communication

In early prehistory, communication was entirely physical: gestures, facial movements, and calls. A hunter's whistle, a protective shout, or a rhythmic chant were rooted in instinct. But when patterns emerged, something changed. The mind began assigning meaning intentionally.

Archaeological and linguistic studies estimate that spoken language may have developed between 50,000 and 100,000 years ago. Not as a single breakthrough, but like erosion forming a canyon, slow, continuous, and shaped by the environment. Mountain regions, coastal settlements, and forest dwellers each developed unique vocal patterns tailored to their world.

The transition from ambiguous cries to structured sounds was one of the greatest cognitive shifts in our evolution.

Words That Outlived Their Speakers

With language, imagination became shareable. A hunter could narrate a strategy so others could improve it. Parents could teach children lessons they never personally experienced. Knowledge stopped disappearing with death. It gained longevity.

Stories, memories, warnings, and beliefs traveled across generations. People could describe not only what they saw but what they hoped for. Language turned experience into legacy.

Once we could speak our thoughts aloud, thought itself began to evolve faster.

How Language Enabled Society

Language did more than help individuals survive. It created us.

With a shared vocabulary, groups coordinated, planned, and formed a culture. Cooperation grew beyond bloodlines. Trust could be built through agreements, promises, and laws. Identity expanded from family to tribe.

Without language:

No shared myths

No moral codes

No collective memory

No civilization

It was the invisible architecture behind every human community.

The First Sunrise of Progress

Before the wheel, the forge, or the first agricultural field, there were words. They allowed humans to think not only in isolation but together. The capacity to communicate across distance and time marked the beginning of conscious collaboration.

Language turned breath into meaning.

Meaning turned curiosity into knowledge.

Knowledge turned survival into progress.

This was humanity's first sunrise, the moment a spark inside one mind could illuminate another.

Chapter 3
Evolution of Storytelling

Storytelling has been part of humanity from the earliest traces of culture. Before ink touched paper and before symbols carved meaning into stone, stories were shared through voice, gesture, and collective memory. They shaped identity, transmitted survival knowledge, and preserved values long before history was recorded. While the formats and tools evolved across eras, the purpose remained steady: to communicate experiences, information, and meaning from one mind to another.

Origins in Oral Tradition

In early human societies, storytelling emerged as an oral practice. Groups shared myths, hunting tales, and ancestral accounts around fires, not as entertainment alone, but as a form of cultural continuity. Without written systems, memory and repetition served as tools for preservation. Bards, elders, and performers carried knowledge across generations through rhythm, emotion, and communal engagement.

Examples appear across cultures, including ancient Greek oral epics, African griot traditions, Aboriginal songlines, and Indigenous American folklore. These narratives were not merely tales but repositories of identity, geography, ethics, and shared history. The audience participated actively, shaping stories with responses, questions, and collective interpretation.

The Emergence of Written Stories

The development of writing created a permanent medium for narrative expression. Civilizations in Mesopotamia, Egypt, China, and Mesoamerica built early writing systems that shifted storytelling from memory to record-keeping. Works such as the Epic of Gilgamesh, early Vedic texts, Egyptian funerary writings, and later, the Iliad and Odyssey, allowed stories to move across distance and generations without relying on oral transfer.

Writing transformed stories into artifacts. Ideas could now travel further than their origin and survive longer than their authors. Knowledge became less vulnerable to memory loss and cultural fragmentation.

Printing and Mass Circulation of Stories

The invention of the printing press in the fifteenth-century Europe expanded storytelling from elite archives to public readership. Books became replicable rather than labor-produced manuscripts. As literacy rates increased, more people gained access to stories, philosophies, scientific discoveries, and religious texts.

The printing press supported cultural shifts such as the Renaissance, Enlightenment, and Industrial Revolution by accelerating the spread of knowledge. Stories no longer depended solely on oral retelling or limited distribution. They became shared cultural assets.

Visual Storytelling and Film

The twentieth century introduced storytelling that combined motion, sound, and visual narrative. The film allowed audiences to see characters move and hear their voices. Stories shifted from imagination to sensory experience.

Silent films evolved into spoken cinema. Black-and-white shifted to color, and later to digital and high-resolution formats. Filmmaking extended storytelling beyond words and into atmosphere, perspective, and collective emotional experience. Global cinema became a cultural mirror, shaping social attitudes and identity in ways written text could not.

Digital Storytelling and Networked Culture

With the rise of the internet, storytelling moved into dynamic and decentralized forms: blogs, online journalism, podcasts, streaming platforms, interactive video, and social-media narratives. Stories became faster to produce and global in reach.

Digital platforms changed not only how stories are told but who tells them. Individuals without institutions now create content that reaches millions. Video games introduced interactive storytelling, where users influence narrative direction. Social platforms turned personal lives into ongoing narratives shared in real time.

Technology did not replace traditional storytelling; it layered new formats onto older structures.

Immersive and Interactive Futures

Today, immersive technologies such as virtual reality, augmented environments, and AI-generated media expand storytelling into experiential spaces. Stories can now be stepped into physically or navigated through choice rather than observation.

While emerging technologies raise questions about authorship, creativity, and authenticity, they continue the same trajectory established thousands of years ago: finding new ways to express meaning through evolving tools.

Continuity Across Ages

Despite shifts in medium from spoken memory to printed pages, screens, and digital worlds, the core function of storytelling remains consistent. Stories help humans interpret experience, share knowledge, connect emotionally, and preserve identity.

Each era did not replace the last; it built upon it. The oral tradition remains in speeches and performances. Written language persists in literature and documentation. Film and digital media extend storytelling into sensory and interactive dimensions.

Storytelling evolves because humans continue to seek better ways to communicate thought.

Chapter 4
Evolution of Art and Painting: The Thought of Visually Expressing Emotions

Art has always been one of humanity's most intimate tools for expressing emotion and recording perception. Long before language matured or societies formalized themselves, people found ways to communicate visually through pigments, carvings, and symbols. Art allowed thought to travel without words, connecting minds separated by time, geography, and culture.

From the first strokes on cave walls to digital canvases, the evolution of art mirrors the evolution of human consciousness. Each era did not simply produce new techniques; it reflected new ways of seeing the world and interpreting our place in it.

Early Expressions of Meaning

Tens of thousands of years ago, in places like Lascaux and Altamira, early humans painted animals, hunts, and scenes of survival. These images were not decorated. They carried emotional weight, possibly ritualistic, cultural, or spiritual. Firelight flickered across the walls, giving movement to painted forms, turning art into experience.

These early works marked the beginnings of visual storytelling. They represented not just memory, but intentional impulse to preserve experience and communicate meaning beyond spoken sound.

Symbol, Power, and Permanence

As civilizations formed, art became a vehicle of identity and authority. In Egypt, stone walls and tombs carried images meant to guide souls through the afterlife. Every figure, gesture, and proportion followed strict rules to preserve religious meaning. Art served eternity more than representation.

Greece shifted focus toward form and idealism. Sculptors carved bodies that captured proportion, symmetry, and motion. Rome added realism, creating mosaics and frescoes that preserved daily life, leadership, and empire.

Art became both documentation and aspiration. It reflected how societies understood themselves.

Faith, Light, and Iconography

Following Rome's decline, religious institutions shaped artistic expression. Cathedrals filled with stained glass and sacred imagery turned narrative into symbol. Artists were often unnamed, their work intended not for personal recognition but spiritual devotion.

In Byzantium and across medieval Europe, images became theological tools bridging believers and the divine. The goal was not realism, but reverence.

Renaissance: Art as Human Expression

In Florence, a shift occurred. The Renaissance renewed attention to human experience, curiosity, and individual vision. Artists became thinkers, engineers, scientists, storytellers.

Leonardo, Michelangelo, and Raphael pushed boundaries of anatomy, perspective, and emotional depth. Techniques such as oil painting introduced new realism and texture, allowing subtle light and shadow to convey psychological nuance.

Art moved from divine representation to human exploration.

Movement, Conflict, and Theatrical Light

The Baroque era introduced intense art filled with motion, contrast, and staged emotion. Caravaggio's shadows, Rembrandt's portraits, and Rubens' dynamic scenes reflected societies wrestling with religion, science, and power.

Later, Rococo softened the tone. Light pastels, romantic themes, and intimate settings reflected a culture shifting toward leisure and ornamentation.

Art adapted to the emotional climate of its time.

Capturing the Moment: Impressionism

By the 19th century, industrial tools like photography disrupted representation. Artists responded by abandoning precision for perception. The Impressionists painted light, time, and sensation rather than form.

Monet's water lilies do not depict a pond; they portray movement, reflection, and atmosphere. Art became experience, not replication.

Breaking the Frame: Modernism and Abstraction

The 20th century fractured aesthetics. War, psychology, and technology changed how artists perceived reality. Picasso's Guernica conveyed the trauma of conflict through distortion. Kandinsky sought to express emotion through shape and color rather than representation.

Art shifted from depicting the external world to mapping the internal one.

Sculpture: Stillness and Form

While painters sought abstraction, sculptors explored expression through mass and silence. Michelangelo's Pietà captures grief in stillness. Rodin's The Thinker embodies contemplation in tension. Sculpture remained grounded in physical presence even when ideas became abstract.

The Digital Renaissance

Modern tools expanded the canvas beyond physical space. Digital media, VR, generative algorithms, and AI enable new forms of expression that are interactive, immersive, and evolving. Art no longer exists only on surfaces; it surrounds, responds, and adapts.

Yet the impulse remains unchanged: to express emotion where language falters.

From cave walls to coded light, art is the journey of turning thought into form.

It is the world reflected through human perception.

Chapter 5
Music and Its Evolution

Music is a universal presence. It crosses geography, belief, and time, moving through cultures with a consistency that few human inventions can match. Whether heard in ritual chants, marching drums, opera halls, or digital streams, music has remained a companion to human life. Long before we carved symbols into cave walls or mapped the stars, we struck stones together, hummed to quiet children, and shaped rhythm into memory. Music did not begin as entertainment. It began as communication, expression, and survival.

Origins: Rhythm Before Language

We cannot pinpoint the exact moment the first musical note sounded. What evidence we have suggests music likely existed before formal language. Early humans developed sound as a way to coordinate hunts, signal danger, or preserve social bonds. Archaeologists have uncovered bone flutes dating back forty thousand years, but the human voice almost certainly predates them as our earliest instrument. It carried emotion when words did not yet exist. Around fires, in caves, and across plains, vocal chants and patterned rhythms formed the earliest collective experiences of sound.

Civilizations and Sacred Sound

As societies formed, music evolved from instinct to ritual. In Mesopotamia, Egypt, and later Greece, music was embedded in ceremony, worship, and state functions. Priests played harps in temples not for performance but for devotion. Greek thinkers treated music as a force that shaped character. Pythagoras observed that lyre strings vibrated in ratios, revealing an underlying mathematical order. His insight did more than influence music; it linked sound to structure and laid the foundations for scientific approaches to harmony.

The Ages of Voices and Instruments

By the Middle Ages, music echoed through stone cathedrals. Gregorian chants, sung in unison, created soundscapes that aimed not to entertain but to elevate consciousness. Outside religious spaces, minstrels and troubadours carried stories across towns, blending narrative, rhythm... social memory. Music was beginning to move from sacred walls into public life.

The Renaissance altered music profoundly. Polyphony allowed multiple lines of melody to coexist, enriching complexity. The madrigal flourished, instruments gained refinement, and the printing press spread music beyond royal courts, expanding access to common households.

The Baroque period pushed emotion to the forefront. Bach, Vivaldi, and Handel built grand compositions filled with contrast and movement. Opera emerged, merging narrative and sound into staged storytelling. The concerto placed individual expression at the center, set against orchestral power.

The Classical period brought structure and clarity. Mozart and Haydn refined formal balance. Beethoven opened the door to emotional depth that foreshadowed Romanticism. The piano's rise offered dynamic range and intimacy, allowing individual composers to express personal voice at scale.

Romantic music pursued emotion with intensity. Orchestras expanded, themes deepened, and music became a direct exploration of personal struggle, national identity, and imagination. Leitmotifs linked sound to meaning, influencing narrative music for decades to come.

Modern Sound and Technological Transformation

The twentieth century broke rules. Stravinsky disrupted rhythm, Schoenberg challenged tonality, and John Cage questioned whether silence could be considered music. Across the Atlantic, jazz grew from improvisation and cultural expression into a global language. Later, rock, hip-hop, reggae, and electronic music formed new sonic identities.

Technology reshaped how music lives. Recording allowed songs to travel without performers. Radio broadcasted broadcasts sound across borders. Vinyl, tapes, CDs, and digital streams each expanded accessibility. Today, music is not limited by instrument or geography. A song recorded in a bedroom can reach millions overnight.

Now, artificial intelligence composes melodies, analyzes patterns, and imitates voices. Virtual concerts expand performance beyond physical venues. These changes raise a new question: what defines authorship when machines can compose?

What Remains Constant

Tools change. Styles evolve. Cultures shift. Yet music continues to reflect what societies feel, fear, celebrate, and dream toward. It is not merely sound. It is a form of thought made audible. It gives shape to emotion and turns inner experience into something shareable.

Music began as vibration and still is. It connects the body, memory, biology, and imagination. In that sense, music remains one of humanity's deepest expressions, not because of how it sounds, but because of how it connects us.

Chapter 6
From Sticks and Stones to Stadiums

Human play predates stadiums, sponsorships, and national flags by millennia. A child tapping a coconut shell in India, or children kicking a bundled sock along a dusty street in Brazil, rarely imagine themselves as participants in history. Yet these modest acts express an instinct deeply embedded in human evolution. They reflect an innate drive to test the body, to compete, to cooperate, and to perform. What appears to be simple play is often a rehearsal for survival.

Early humans depended on agility and coordination to hunt and to defend their communities. Chasing prey, throwing stones with precision, and wrestling for dominance were not recreational diversions. They were essential practices that refined the skills required for endurance and survival.

Archaeological evidence from ancient Egypt, Mesopotamia, and Greece reveals organised wrestling, racing, and team contests. Communities gathered as spectators. Skill became performance. Performance became tradition. In this transition, play evolved beyond physical practice and emerged as culture.

Competition Becomes Ceremony

In 776 BCE, ancient Olympia hosted the first recorded Olympic Games. City-states that often clashed militarily paused their conflicts to compete peacefully. The arena replaced the battlefield. Victory became honor rather than survival. The Olympic flame symbolized unity, reminding participants that humans could challenge each other without destroying each other.

Rome later expanded this instinct into a large-scale spectacle. The Colosseum is filled with crowds seeking drama and control. Gladiators fought for survival. Rulers used sport to display power. Even in brutality, one idea persisted. Spectatorship shapes society. Sport became a mirror of human values.

From Villages to Institutions

During medieval Europe, people continued to compete in jousting, archery, and communal games. But organized sport, as understood today, emerged more recently.

In the nineteenth century:

Rules were written

Uniforms and referees created fairness

Schools integrated sports to teach discipline and teamwork

Play evolved into an institution. Nations formed sporting identities.

In 1896, Pierre de Coubertin revived the Olympic movement. With the motto faster, higher, stronger, global sport gained a new mission. Unity through performance. What began in ancient Greece now gathers the world every four years.

Sport Shapes Identity

As games spread across continents, entire cultures formed around them.

Cricket became a shared heartbeat in India and Pakistan

Football grew into the emotional language of Argentina, Brazil, and Europe

Basketball told stories of city identity and aspiration across the United States

Each sport built its own world. Its own heroes. Its own history.

Technology accelerated this growth. Cameras brought matches into homes. Slow-motion replay refined judgment. Wearable devices and analytics transformed performance. Coaching became a science.

By the late twentieth century, the stadium was no longer the only arena. The screen became another field. Fans gathered online. Geography stopped limiting participation.

The Digital Revolution in Play

In the 1990s and early 2000s:

Online communities formed global fanbases

Fantasy leagues turned spectators into strategists

Live streaming opened doors to international audiences

Then, social media shifted the balance of attention. Athletes could speak directly to the world. Influence extended beyond the scoreline.

New technologies continue to reshape competition:

Artificial intelligence helps scouts discover talent

Virtual reality places fans inside the game

Blockchain reimagines tickets and collectibles

Esports emerged with arenas, coaches, professional teams, and worldwide tournaments. The keyboard became a new form of equipment. Reaction time replaced muscle as the deciding force.

What Remains Constant

Despite every transformation, a core truth holds steady. Sport is effort. Rivalry. Collaboration. Resilience. It is a test of the human body against its own potential.

When Usain Bolt accelerates beyond history, when Serena Williams redefines strength and precision, or when a child kicks a plastic bottle with a dream in their chest, the same instinct lives on. The desire to strive.

Beneath bright stadium floodlights or under a warm sun on a quiet dirt road, sport unites us. Through movement. Through emotion. Through aspiration.

Play reveals who we are. And who we hope to become.

Chapter 7
The Symphony of Vibrations – How We Listen to the World

In this world, everything moves. From atoms shifting in silence to thunderclouds rolling across the sky, nothing is ever truly still. The universe communicates through pulses, frequencies, and patterns, and we perceive this ceaseless motion through a remarkable biological gateway: the ear.

We do not merely hear sound. We perceive vibration.

Sound is not an object or a substance. It is energy traveling through a medium such as air, water, or solid matter. When a voice rises in conversation, or a guitarist strikes a string, waves radiate outward in patterns too rapid for the eye to detect. These waves continue their journey until they reach the outer ear, where they are guided toward the eardrum.

Here, vibration becomes perception.

The eardrum moves in synchrony with these waves, transmitting delicate tremors through the three tiny bones in the middle ear. These bones amplify the motion and deliver it to the cochlea, where thousands of microscopic hair cells convert vibration into electrical signals. Only then does the brain construct what we recognize as sound.

Without motion, there is only silence. Without vibration, the world would have no voice.

This principle governs the music we cherish, the voices we trust, and the environments we inhabit. A piano produces music not because a key is pressed, but because its internal strings vibrate at precise frequencies. A singer's voice is not merely an idea; it is a vibration shaped by breath, muscle, and resonance. Even the gentle rustle of trees at dusk or the distant surge of waves is energy in motion, traveling until it encounters a body capable of receiving it.

The world vibrates. We resonate.

If the eardrum were unable to move, even the loudest applause or the deepest bass note would pass unnoticed. Sound exists not simply because it is generated, but because our bodies are capable of responding to it. Hearing is, at its core, synchronization, a shared rhythm between environment and biology.

Scientists suggest that vibration may extend far beyond sound alone. According to theories in physics, including string theory, the fundamental structure of matter itself may consist not of static particles, but of minute vibrating strands of energy. If this is true, motion is not merely a feature of sound; it is the underlying architecture of reality.

The next time you listen to a song, what moves you is not only melody or lyrics, but the physical waves traveling through the air and activating your nervous system. Your body becomes part of the composition. You are not observing sound; you are participating in it.

This realization carries a profound implication. Sound is not external to us; we are embedded within the motion that generates meaning. We do not simply receive the world. We are joined to it.

From Listening to Speaking: Vibration as Communication

Understanding vibration does more than explain how we hear; it reveals how we influence others. Every voice carries not only words, but frequency, tone, and presence. People respond not simply to what we say, but to how our vibration reaches them: confidence, calmness, hesitation, authority, or tension.

A powerful speaker is not defined by volume, but by resonance.

Leaders, orators, and performers shape air into emotion, directing vibration with intention. The diaphragm becomes an engine, the chest a chamber, and the skull a resonator. Speech is not merely language; it is physics expressed through biology.

To speak boldly is to control vibration.

To persuade is to align emotion with sound.

To inspire is to send waves strong enough to move another mind.

Whether addressing a crowd or whispering to a friend, the voice is a tool of influence, and like any instrument, it strengthens with awareness and practice.

We Are Instruments of the World

The universe does not speak in silence. It speaks in waves. From the shape of galaxies to the pulse of music, from a whispered thought to the roar of an engine, everything around us moves in patterns of vibration.

We are part of that movement, not observers, but participants.

We do not merely listen to the world.

We resonate with it.

And sometimes, when we speak, we become part of the music that shapes others.

We are not just listeners of sound.

We are instruments of it.

Chapter 8
The Peacock's Feather – Where Light Learns to Dance

I once watched a peacock, not for its movement or call, but for its colors. A single feather lay still, untouched by pigment or dye, yet the moment light touched it, a cascade of blue, green, gold, and violet emerged. It was not painted. It was not stained. And yet it shimmered as if light itself had been sculpted into color.

It may appear mystical, but it is not. It is science. Silent, intricate, and formed through millions of years of evolution.

For centuries, we tried to recreate the peacock's brilliance using paints and dyes. But the bird's tail does not contain blue or green pigments. The color does not arise from chemical composition. It arises from structure.

Under a microscope, the feather reveals microscopic rods of melanin embedded in keratin. These rods are not randomly scattered; they are arranged in a precise geometric lattice, a natural version of a two-dimensional photonic crystal. Evolution, not human engineering, designed this nanoscale system long before science understood it.

When light hits this structure, wavelengths interact. Some frequencies are suppressed; others are amplified. The result is structural color, produced not by chemical absorption but by controlled interference. Color becomes a function of architecture.

The peacock does not wear color. It constructs it.

As scientists observed this mechanism, a question followed: could humans build color the same way? Could we create hues that never fade, because they are created not from pigment but from engineered structure?

This led to the rise of biomimetic photonic engineering.

Researchers at MIT, Harvard, and European laboratories studied feathers, butterfly wings, and beetle shells to replicate their nanoscale designs. These natural templates inspired technologies ranging from anti-counterfeit holograms to color-shifting textiles and energy-efficient displays.

Companies such as Opalux and Qualcomm worked on e-ink screens that remain visible in direct sunlight while consuming minimal power. Cosmetic researchers explored makeup that shifts shades through structural reflection rather than dyes.

All from a single principle found in a feather.

Structural color challenges an old assumption: that color must be applied. Instead, it can be revealed. It suggests that color is not merely a surface property, but an interaction between matter and light.

The peacock offers a quiet lesson. Nature does not decorate. It designs.

Chapter 9
The Creature Without Bones That Taught Us to Build Robots

Imagine a machine with no rigid skeleton, no hinges, and no wheels, yet it glides, curls, grips, and adapts with remarkable precision. It can fit through a gap no wider than a coin, hold an egg without cracking it, and navigate places where traditional mechanical systems would jam or fracture. This is not science fiction. It is a design borrowed from a creature that has been refining its mobility for more than three hundred million years.

The octopus.

Eight arms. No bones. A body that functions entirely through fluid motion. In many ways, it represents an evolutionary alternative to the rigid structures that define most animal bodies, including our own.

In the natural world, the octopus is a paradox. It has no internal or external skeleton, yet its limbs move with intention and force. Its muscles work like biological hydraulics, stiffening when required and softening when relaxed. Every section of each arm can twist, lengthen, bend, or contract independently. In a sense, the octopus is not a body guided by a single central brain, but a distributed system in which each arm acts with a degree of local control.

To engineers raised on a century of metal frames, hinges, and motors, such a design once seemed impossible to replicate. But to a new wave of roboticists, it became a blueprint for the next generation of machines.

In 2015, teams at Harvard's Wyss Institute and Italy's Istituto Italiano di Tecnologia began studying octopus biomechanics to answer a pressing question: how do we build robots that do not require skeletons at all? Machines that resemble living organisms not only in appearance, but also in flexibility, sensitivity, and environmental adaptability.

This was the birth of soft robotics.

Instead of steel, carbon fiber, or servo motors, engineers began experimenting with silicone-based materials, pneumatic actuators, and polymer structures that behave like living tissue. These machines did not roll or clatter. They slid, compressed, stretched, and flowed.

The ocean became the first testing ground. Octopus-inspired robotic arms were developed to perform delicate underwater work, from handling coral samples to neutralizing explosive devices without damaging surrounding ecosystems.

Medicine soon followed. Engineers began designing endoscopic devices that can move gently through the human body, navigating narrow pathways while adjusting shape during surgery. Unlike rigid tools, these systems can change form rather than forcing tissue aside.

One prototype, known as the "Octobot," was built without electronic circuitry. It relied on fluidic logic systems powered by pressurized gas traveling through soft channels, echoing the decentralized communication found in octopus arms. This opened possibilities for machines that operate where traditional motors and wiring would fail.

Even space agencies are paying attention. NASA has explored soft robotic systems for missions to icy moons such as Europa, where rigid rovers might fracture under shifting terrain. A robot that flows rather than walks may be better suited to alien landscapes.

It is surprising to realize that the future of robotics may not come from steel, precision gears, or rigid frames, but from a creature that drifts silently through the sea.

The octopus never intended to teach us anything. Yet by simply existing, adapting, and surviving through evolutionary time, it provided a blueprint for machines capable of bending, sensing, and responding more like life itself.

Softness, often dismissed as weakness, is becoming a new kind of strength.

Chapter 10
Ocean: The Mysterious Blue That Drew Us In

I sit by the shore, watching waves rise, fold, and fall in a rhythm older than memory. The ocean meets the sky in a straight line that seems endless. For as long as humans have existed, this vast blue space has called out to us. It was not simply a limit. It was a question that kept returning. What lies beyond this water? What waits out there?

Even thousands of years ago, when humans had no tools for navigation, they still stood at coastlines and felt that same pull. The sea offered clues. Salt in the wind, tides that behaved like breath, shells and fish delivered to the shore. Life existed within that water. Where life exists, humans find purpose, food, movement, trade, and curiosity.

Our journey onto the ocean did not happen at once. It was a slow rising of thought, shaped by fear, imagination, hunger, and courage working together.

From Floating Logs to Navigating Stars

Perhaps the first voyages were accidents. A person holding onto a drifting log after a storm. A family carried away by a river flood. But humans do not stop at accidents. We repeat discoveries. We improve them.

Around ten thousand years ago in Southeast Asia, people began tying logs into rafts. Later, they carved canoes with sharper shapes that cut through water. Knowledge passed quietly through generations. Each improvement was a memory kept alive by experience.

Then sailors learned to use the wind. A simple cloth raised upright could catch air. With that discovery, humans no longer waited for the sea to decide their direction. They began choosing their own. Polynesian navigators became the masters of this new world of movement. They travelled across the Pacific Ocean, guided only by stars, cloud shapes, the behaviour of birds, and the feeling of waves beneath the hull. They carried entire maps in their minds.

Human thought became a compass.

Ocean as Food, Trade, and Belief

In many places, people first entered the sea to eat. Fish migrated in predictable patterns. Tides revealed oysters and crabs. Early societies built hooks, nets, and spears. In some regions, divers held their breath for long minutes to bring food up from the seafloor. The Ama women of Japan have done this for nearly two thousand years.

Fishing was more than survival. It became tradition, trade, and often spirituality. Many coastal communities believed the sea was a living presence, a giver that deserved respect. The question was not only how much could be taken but also how to continue receiving without creating harm.

As skills improved, the sea turned into a wide road for goods. The Mediterranean became a centre of exchange. Egyptians, Phoenicians, Greeks, and later Romans traded metals, glass, spices, and ideas. Long before Europe understood its value, the Indian Ocean connected East Africa, Arabia, India, and Southeast Asia in a busy network of merchants and sailors.

Objects travelled. Stories travelled. People are changed by meeting one another. A single pot from China could journey to the coast of Africa. Seas connected cultures that might never have met otherwise.

Discovery and Conquest

In the fifteenth century, powerful European kingdoms pushed into the ocean with a new intention. They wanted control. Explorers and their rulers funded long voyages to find faster routes to wealth. Voyages by Columbus, da Gama, and Magellan changed the world map.

But curiosity brought consequences. Lands already filled with people were renamed. Millions were taken from their homes and forced across the Atlantic as enslaved labour. Oceans became tools for conquest, not connection.

Human thought influenced the ocean. Now, human greed shaped it as well.

Return to Curiosity: Science and Protection

In recent centuries, our relationship with the sea has shifted again. We turned downward instead of outward. Submarines travelled into deep darkness where strange creatures live without sunlight. Scientists discovered underwater mountains taller than anything on land. Cameras showed coral cities filled with life.

We learned that the ocean controls Earth's climate. It stores heat. It produces oxygen. It feeds billions of people.

Yet as our knowledge grows, so does the problem. Overfishing, pollution, and warming waters threaten the balance that supports life. The new challenge is responsibility. We must continue learning without destroying what we depend on.

The Same Call

Our vessels have changed. Rafts became ships made of steel. Star maps became satellites. Curiosity became calculation. Despite all this progress, most of the ocean is still unexplored.

When we look toward the horizon today, a familiar thought remains inside us. It is the same thought our ancestors felt while standing at the edge of the shore:

There is more.

Go and understand it.

Chapter 11
The Thread of Genius

When Humans Looked at the Spider and Saw the Future

It never begins with thunder. Human breakthroughs often start as quiet curiosity. Someone paused to watch a spider. A small creature is constructing its web with precision most architects would envy. A single line of silk that looked fragile to the eye yet held extraordinary strength. How could something so thin, so delicate, contain such remarkable power?

From that moment of attention, an idea began to form.

Spiders have been spinning silk for more than 400 million years. Humans, however, only recently began asking deeper questions. Could we understand this natural engineering? Could we recreate it? Could the spider's genius become ours?

That brief observation of a web suspended between two branches would eventually influence laboratories, industries, medicine, and even military innovation. A simple wonder evolved into a chain of discoveries.

The Strength in Fragility

Spider silk is not just beautiful. It is one of nature's most advanced materials. When compared by weight, it is stronger than steel and tougher than Kevlar. Scientists soon realized it was not only about raw strength. It was the versatility that amazed them.

Spiders spin several kinds of silk.

Some sticky, some smooth.

Some elastic, some rigid.

Some meant for catching prey, others for protecting new life.

Nature designed a toolkit of fiber solutions inside a single tiny creature.

Its secret lies in its molecular structure. Highly ordered crystals are supported by flexible protein chains. This gives silk both impressive tensile strength and surprising elasticity, a combination engineers struggle to achieve.

One scientific demonstration often captures attention. A pencil-thin bundle of spider silk, roughly one centimeter in thickness, could theoretically lift the Statue of Liberty. The tensile strength of certain species reaches more than one gigapascal. That means hundreds of metric tons of weight held by a material lighter than cotton thread.

Yet we must be honest. These comparisons are theoretical. Spiders do not produce silk in large quantities. They are solitary hunters and cannot be farmed together because they will prey on each other. Even the most ambitious projects using real spiders could collect only small amounts of silk.

Still, this discovery shifted our mindset. Instead of seeing spiders as ordinary, we began to recognize that nature often solves engineering challenges long before we do.

From Web to Wonder

Once we understood the limits of harvesting natural silk, a new idea emerged. If we cannot farm spiders, perhaps we can produce the silk proteins another way.

That thought led scientists to goats.

By inserting genes from spiders into goats, the modified animals produced silk proteins in their milk. These proteins could then be collected and spun into what became known as Bio Steel. It was not imitation but biological collaboration: the code of a spider expressed through a mammal.

Defense researchers quickly became interested. Materials that are strong, flexible, and lightweight are essential for protective gear. Could spider silk become the future of body armor?

The image of a soldier protected by a material inspired by a spider may feel poetic. Yet it represents the intersection of science, biology, and imagination.

Medical Miracles in the Web's Wake

Doctors and scientists began to see silk as more than armor.

Spider silk is biocompatible. The body accepts it rather than rejecting it. This opened a new path for medicine.

Researchers explored silk as a scaffold to help nerves regenerate.

Surgeons used silk sutures that dissolve gently over time.

Material scientists experimented with silk carriers for targeted drug delivery.

At Tufts University, studies showed that silk could replace damaged ligaments and tendons. It resists bacterial growth and naturally degrades, which reduces complications.

Nature had already written the instructions. We simply learned to follow them.

Fashion Caught in the Web

In 2012, at the Victoria and Albert Museum in London, a cape made entirely from golden orb-weaver silk astonished visitors. More than one million spiders were carefully milked to create it. The garment was not a product of mass manufacturing. It was a tribute. A reminder that nature's engineering can also be art.

Innovation did not stop at admiration. Bolt Threads, a biotechnology company, developed Micro Silk by growing spider silk proteins through yeast fermentation. Clothing made from this fiber offered both style and sustainability. Materials of the future may not come from fields or factories but from biological processes shaped by evolution.

Thought in the Shape of a Web

The deeper scientists looked, the more they realized that the spider is not simply a spinner of silk. It is an engineer, a chemist, a builder, and a problem solver.

NASA sent spiders to space in 2011 to observe how they behaved in zero gravity. Disoriented at first, they eventually adapted and formed webs with new geometry. Even in unfamiliar environments, instinct guided them to redesign. The web remained functional.

There is a lesson in that. Intelligence takes many forms. Some of it is encoded in behavior, not words.

A Thought that Keeps Spinning

What began as a quiet moment of observation has expanded into industries and innovations. That is the essence of biomimicry. We begin with humility. We watch carefully. Then we learn.

Inside every strand of engineered silk is a reminder. A spider once did this without machines or theories. Nature often provides solutions long before we know we need them.

To progress, we do not always need to create new ideas from nothing. Sometimes we only need to notice what has been here all along.

We look. We learn. And then we think.

Chapter 12
When Humans Decided to Clothe Themselves

I run my hand across the fabric of my shirt, and a curious thought arrives. *What if* no human ever imagined thread? Would we still be wrapping ourselves in leaves or animal skins? Somewhere in prehistory, a human looked at fibers from a plant or wool from an animal and imagined a new possibility: twist, spin, weave. That small leap of thought transformed how we live.

We do not know the name of the first weaver. There is no monument for the person who discovered that twisted fibers become thread and that thread becomes cloth. Yet we can imagine that thinker. Perhaps seated by a riverbank, touching flax strands, watching the wind, seeing patterns in nature, and daring to copy them. Clothing began as protection, yet soon it became more than warmth. People wanted expression and identity. A hide covers the body. Clothes speak.

This idea, to transform natural fibers into something useful and also beautiful, became a foundation of civilization. Egyptians refined linen from flax thousands of years ago. Priests wore the purest linen to symbolize spirituality and order. In the Indus Valley around 3000 BCE, farmers cultivated cotton and spun it by hand, producing finely woven garments that travelled along early trade routes. In China, observers studied a silkworm spinning its cocoon and saw wealth in a single filament. Silk became a luxury that reshaped global exchange.

Each civilization adopted its own fabric traditions. Greeks used wool from hillside flocks. The Romans expanded trade networks so they could acquire rare dyes and fabrics. Purple pigment, extracted from countless sea snails, became a signal of royal authority. Clothing created hierarchy. Cloth could separate rulers from subjects, rich from poor, purity from impurity.

The loom appeared and turned weaving into a rhythmic skill performed daily. Homes echoed with the pattern of pull and release as threads became textiles. Mothers taught daughters. Knowledge passed hand to hand. A community's story was stitched into every garment.

Human ambition rarely rests. At some point, a spinner looked at the slow-turning spindle and imagined a faster rhythm. That impulse changed the world.

The eighteenth century introduced machines into the quiet world of weaving. The spinning jenny allowed multiple threads to be spun at once. Richard Arkwright improved the process with water power. Edmund Cartwright followed with a mechanical loom that transformed speed and scale. In only a few decades, weaving shifted from homes to factories. Fabric production increased dramatically. Clothing became more accessible. A craft that began in small villages helped ignite the Industrial Revolution.

Innovation continued to evolve. In 1935, nylon was created in a laboratory in the United States. It was the first synthetic fiber produced entirely by human design. Cloth no longer depended on plants or animals. It no longer rotted, wrinkled, or soaked easily. This was a new form of power, a direct reshaping of nature.

By the late twentieth century, clothing became a language of identity. Jeans, once worn for labor, turned into symbols of rebellion and culture. Suits became markers of status. Traditional garments like the sari, kimono, and hijab carried longer stories: faith, ancestry, and community. Clothing became emotion made visible.

Today, the landscape of fashion is still changing. Factories produce clothing in enormous quantities. Designs created by artificial intelligence appear instantly on global platforms. Scientists are developing fabrics that monitor heart rate, control temperature, and biodegrade safely. In laboratories, spider silk proteins are engineered for strength and flexibility. We even print clothing using digital technology.

Yet despite all these advances, we continue to reach for cotton, linen, and silk. They connect us to human beginnings.

A shirt worn today may include threads sourced from several continents, woven by machines, and guided by software. Still, the core thought behind it remains ancient. A human once looked at a plant stalk or an animal's wool and imagined a future stitched together from thread.

Clothing did more than protect our skin. It expanded my imagination. It allowed us to create beauty, status, memory, and meaning. In that discovery lies one of humanity's quiet but powerful achievements. Not only the ability to survive, but the courage to create.

Chapter 13
The Thought That Froze Time: From Clay to Paper

A soft breeze enters the room as I guide ink across a sheet of paper. Each letter becomes a shape that represents something invisible, a thought transforming into form. I pause. The page feels weightless beneath my hand. Yet this simple action stands at the end of a journey that began more than twenty thousand years ago. A journey not powered by conquest, but by the need to preserve the human mind.

Everything started in silence.

Spoken words vanished the moment they were heard. Language gives thoughts a voice, but a voice disappears. Humanity discovered the truth. Ideas needed a container. Something that could hold meaning for more than a moment. Something that could survive the speaker.

The earliest answer appeared as marks carved into bone and pressed into clay. The answer was writing.

Marks on Bone, Echoes in Time

Long before alphabets existed, humans left behind tally marks on bones and stones. These silent scratches recorded quantities, lunar cycles, or perhaps early measurements. One famous example is the Ishango bone from Central Africa, dated around 20000 BCE. It carries notches that may reveal arithmetic or timekeeping.

These markings were the earliest footprints of thought. Evidence that ideas no longer needed a voice to live.

However, society expanded. People formed groups, traded goods, and made promises. Life demanded a better way to remember. Not just numbers, but agreements. Not just quantities, but history.

Clay: The First Library

Around 3400 BCE in Mesopotamia, scribes pressed symbols into wet clay using reeds. These impressions formed cuneiform, one of the first writing systems. At first, it recorded economic transactions such as grain, labor, and livestock. Soon, it recorded beliefs, stories, and laws.

The Epic of Gilgamesh, the oldest surviving literary work, was written on clay tablets. These tablets lasted for thousands of years. They were strong, but they were also heavy. A single record could weigh as much as a small stone.

Humanity had learned to write, but thought still carried physical weight.

Papyrus: When Ideas Learned to Travel

Around 3000 BCE in Egypt, people sliced papyrus plants into thin strips and pressed them into sheet form. The result was light and flexible. Knowledge could now move. Scrolls slipped into saddlebags or were hidden in robes. Messages could cross deserts and seas.

Papyrus inspired a new ambition. Rather than storing only accounts, people began collecting knowledge itself. In Assyria, King Ashurbanipal filled great rooms with tablets. Later, the Library of Alexandria became a beacon of written wisdom. Thousands of scrolls rested in silence, waiting for hands from distant times.

These scrolls were more than records. They were voices, preserved.

Paper: When Thought Became Infinite

While papyrus reshaped writing in the West, something even more transformative emerged in China. Around 105 CE, during the Han Dynasty, a court official named Cai Lun created the first true paper by pulping plant fibers, rags, and fishing nets.

Paper changed everything. It was easier to produce and easy to fold, stack, and bind. Writing no longer belonged only to scribes or rulers. Merchants, monks, and travelers could now carry ideas with them. The written word began to spread faster than empires.

Thought had become a seed that could travel across borders and survive beyond lives.

The Real Invention Was Preservation

From tally marks in caves to the rustle of printed pages, humanity's mission remained constant. We wanted to make our thoughts stay. We wanted to hold an idea still. We wanted memory to outlive the body.

As I watch the ink dry on the page in front of me, I feel connected to a lineage of creators who refused to let knowledge fade. Mesopotamian record keepers, Egyptian papermakers, Chinese artisans, and medieval librarians all share a single purpose.

We preserve thought so it can continue.

Today, new questions arise. Will we write without hands? Will ideas exist as data inside clouds and circuits? Will memories return as easily as a file retrieved?

Perhaps that future is near. Yet for now, the gentle friction of ink on paper and the quiet permanence of the written word still carry their own power.

Thought remains alive each time we read.

Thought remains alive each time we write.

Chapter 14
The Man Who Gave Thought a Medium

It is easy to forget the power of a sheet of paper. A rough note in a pocket. A child's artwork taped to a wall. A page turning under a curious finger. Yet behind its familiar softness lies one of the greatest turning points in human history. A turning point was sparked not by a king or conquering hero, but by a quiet court official in ancient China.

His name was Cai Lun. He gave thought a home. He allowed ideas to travel.

The year was 105 CE, during the Han Dynasty. Cai Lun worked in the imperial workshops of Emperor He, a place where tools of ceremony, ink, and military equipment were crafted with precision. But his imagination did not cling to metals or jewels. It reached for something more essential. Something that could move knowledge beyond the walls of palaces.

He saw the world differently. Records carved on bamboo sticks were heavy and difficult to store. Silk, soft and elegant, was far too valuable for common writing. What civilization needed was a surface that could belong to everyone.

He began to experiment. Tree bark. Strips of hemp. Worn rags. Old fishing nets that no longer held the sea. He soaked these discarded materials. He pounded them into fine pulp. He stirred the mixture through water, then lifted it on a flat screen. When the water drained away, a thin layer of fibers remained. Pressed and dried, it formed something light. Something smooth. Something new.

Paper.

Not luxury. Not a burden. Possibility.

When he presented his creation to the emperor, it was immediately recognized as a leap forward for the empire's intelligence and record-keeping. Imperial scribes could write faster. Messages could travel farther. Knowledge could be stored without weighing down the shelves of the world. History would later record his contribution in the Book of Later Han. A quiet recognition of a civilization transformed.

Paper did not stay where it was born. Along trade routes, it traveled with merchants, monks, soldiers, and stories. In 751 CE, papermakers carried their craft westward into Samarkand. From there, the technique reached the Islamic world and then Europe. When Gutenberg's printing press multiplied words for millions, it stood on the foundation Cai Lun had placed centuries earlier. His idea became global before the word global existed.

Yet while his invention soared, life at court remained unpredictable. Politics shifted. Alliances changed. Records suggest that Cai Lun faced accusations and pressure that darkened his final days. His story ended quietly. But his legacy never did.

The paper preserved the thoughts of emperors and farmers, scientists and poets, dreamers and dissenters. It gave laws longevity. It turned fragile memories into durable history. It carried love letters, revolutions, mathematics, and the very idea of progress.

Today we write. We print. We publish. We learn. And every page that turns carries a fragment of his curiosity.

In a world overflowing with ideas, a man once looked at a pulp of waste and imagined knowledge taking shape. His creation did not just store information. It allowed thought to multiply, far beyond the limits of one lifetime.

Cai Lun opened a doorway. We walked through it.

Chapter 15
Printing — When Thought Multiplied

Ideas first lived in fragile places.

They were scratched into clay, brushed onto papyrus, or copied by tired hands onto delicate sheets of paper. Knowledge existed, but it travelled slowly. Every copied book demanded hours of human effort. Thought moved, yet it moved like a candle flame struggling against the wind, always at risk of vanishing.

Then history changed.

In the mid-15th century in Mainz, Germany, a goldsmith named Johannes Gutenberg combined several existing technologies into one world-shifting machine. Movable metal type, oil-based ink, and a screw press came together through his ingenuity. In 1455, the first copies of the Gutenberg Bible appeared. But this moment was bigger than a single book. It marked the birth of mass communication.

What once took monks months or years could now be produced in days.

The guardianship of knowledge began to shift.

Scribes, nobles, and clergy could no longer control information alone.

Books reached the hands of merchants, artisans, and ordinary families.

Human thought gained freedom.

The printing press did not simply preserve ideas.

It accelerated them.

Martin Luther's Ninety-Five Theses spread across Europe because thousands of printed copies carried his challenge far beyond one church door. His message ignited the Protestants.

Reformation.

In the same era, Copernicus quietly published his groundbreaking work that placed the sun, not Earth, at the center of the cosmos. The printed word helped fuel the Scientific Revolution that followed.

As more books reached more minds, imagination widened.

Chaucer, Dante, Shakespeare, and countless others found audiences far beyond royal courts or cathedral libraries. Their stories and ideas lingered in taverns, classrooms, and public squares.

By the 17th century, newspapers emerged as thin sheets carrying the heartbeat of nations. Revolts, discoveries, and daily events travelled faster than ever. A reading culture began to rise.

The Industrial Revolution of the 19th century only accelerated this transformation.

Steam-powered presses produced stories and news at an unprecedented scale.

Reading evolved from privilege to routine.

Literacy expanded.

Curiosity spread.

Yet the evolution of reading did not stop with paper.

In the 20th century, a quiet transition began with the first digital texts. Early e-books existed mainly in universities and laboratories during the 1970s. The change reached everyday life decades later, especially with devices like the Amazon Kindle in the 2000s. Entire libraries became accessible inside a single lightweight screen.

Soon after came audiobooks.

From tapes and compact discs to streaming apps, the voice returned as a vehicle for stories. Humanity rediscovered the oldest form of storytelling: listening. But now we listen while walking, cooking, traveling, or at work. Reading extended beyond the eye. Thought became portable.

From clay tablets to printed scripture, from ink-covered presses to digital files, the purpose has remained steady across centuries. Humans record ideas so they can survive, travel, and spark new ideas in others.

Now we stand at the edge of yet another shift.

Artificial intelligence writes and analyzes text.

Technology experiments with interfaces that may one day connect human thoughts directly to written language.

Meanwhile, one constant remains:

Every innovation in reading is driven by the same desire that sparked the first marks on clay.

We write so we can think together.

We read so we can remember.

And we continue building new tools so our ideas can travel further than any single mind.

The form will change.

The mission will not.

Humanity must keep reading.

Chapter 16
Eyewear – The Thought That Helped Humanity See Better

I sit near a window, watching sunlight ripple across leaves while my eyes adjust without effort. It is remarkable how the lens, retina, and brain cooperate to make vision feel effortless. Yet nature, for all its brilliance, does not always offer perfection. When age distorts clarity, or when someone is born with a blurred world, what do we do? Who first imagined a tool that could restore sight?

Perhaps it began with a simple struggle. An elderly scholar leans closer to decipher a fading manuscript. A craft worker squinting to complete a delicate design. In those quiet moments, someone must have thought, If only I could make the world sharper again.

Polished Stones and the Curiosity of Vision

The earliest visual aids did not resemble modern glasses. They started as polished stones used to magnify objects. By the first century CE, people were experimenting with glass filled with water to enlarge text. Historical accounts suggest that the philosopher Seneca could read small writing by viewing it through a glass globe of water.

Small discoveries continued to build upon one another. During the Arab Golden Age, Ibn al-Haytham, known in Europe as Alhazen, transformed curiosity into science. His Book of Optics in the 11th century explained how light travels, how the eye sees, and how lenses could be shaped for magnification. His ideas influenced European craftsmen centuries later, guiding them to use lenses not only to enlarge but to correct.

The Venetian Spark and the Birth of Spectacles

By the late 13th century, artisans in Italy began creating convex lenses designed to improve near vision. These early spectacles did not sit comfortably and had no side arms. They were held by hand or balanced on the nose, yet they changed lives.

A Dominican friar, Giordano da Pisa, publicly acknowledged the new invention. He wrote that one could now read with the help of glasses even when the eyes had weakened. This statement marks one of the earliest records of spectacles being widely recognized.

The idea behind them was revolutionary. Poor vision was no longer a condition that must be accepted without hope.

Vision Evolves Across Centuries

In the 15th century, concave lenses were introduced for people who struggled to see distant objects. The printing press soon followed, and as books multiplied, so too did the need for reading glasses.

By the 17th century, spectacles had become common tools for daily life. Edward Scarlett, a British optician, added side arms in the 1720s so glasses could rest comfortably on the ears. Comfort and stability became part of the design.

Later, in the 18th century, Benjamin Franklin faced the challenge of constantly switching between glasses for reading and for distance. His solution was simple and brilliant. He combined two lenses into one pair of frames, creating the first bifocals. A practical invention born directly from lived experience.

The Industrial Boom and Eyewear for Everyone

In the 19th century, technology changed everything. Eyewear moved from small workshops into factories. Glass became clearer and more precise. Metal frames replaced bone and wood. Ordinary workers could now afford vision correction, not just scholars or wealthy elites.

In the 20th century, glasses evolved into a personal style. Frames became lighter and more expressive. Eyewear became an identity. Malcolm X's bold black frames and John Lennon's round spectacles showed how glasses could represent personality as much as purpose.

Lenses That Disappeared and Returned as Contact

The mid 20th century introduced contact lenses. Early versions were rigid and uncomfortable, yet they allowed vision correction directly on the eye. By the 1970s, soft hydrogel lenses arrived, offering comfort and freedom. People could now see clearly without adjusting frames or cleaning fogged lenses.

Daily disposables and multifocal contacts later expanded choice and convenience. Seeing better no longer required wearing something visible.

The Digital Age and Smart Eyewear

Screens dominate modern life. Eyes strain easily. Blue light filtering lenses were created to reduce fatigue. Technology pushed even further and turned eyewear into an interactive device.

Smart glasses emerged, capable of displaying maps, taking photographs, and translating speech. Eyewear no longer only improved sight. It improved interaction with the world.

Augmented reality added another layer. Information could appear in real space while someone walked, worked, or explored. Glasses became gateways to data.

The Future of Seeing

Tomorrow's eyewear may monitor health, adjust focus automatically, or scan for early signs of disease. Artificial intelligence could personalize vision in real time. Glasses may quietly protect us from dangers we cannot see.

We started with reading stones. Now we imagine glasses that can think. What remains unchanged is the reason behind every improvement. Humans strive for clarity, for better vision both physically and intellectually.

Eyewear reflects a continuous pursuit. From magnifying stones to intelligent lenses, the thought has always been the same. We want to see, understand, and navigate the world with greater confidence and insight.

Chapter 17
The Instruments of Discovery: Telescopes and Microscopes

The night sky stretches above us, a dark canvas scattered with stars that shine like distant promises. Beneath this vastness, I often pause and ask myself what countless humans before me have also wondered. What exists beyond that darkness. What secrets hide where our eyes cannot reach. This instinct to go further, to understand more than what is visible, has shaped our history. Two inventions changed everything, revealing worlds both unimaginably distant and unbelievably small. The telescope and the microscope.

The Telescope: Reaching for the Stars

In 1608, a spectacle maker in the Netherlands named Hans Lippershey was arranging lenses when he noticed something unexpected. When aligned in a particular way, faraway objects suddenly appeared closer. It was not a planned discovery. Yet from that moment, the sky would never look the same again.

Word of the invention reached Italy. In 1609, Galileo Galilei began improving its design. He built a telescope that magnified twenty times more than the naked eye. When Galileo turned his instrument toward the sky, what he saw challenged centuries of confident belief. The Moon was rough and mountainous instead of perfectly smooth. Jupiter had moons of its own that orbited in direct conflict with the idea that everything revolved around Earth. Venus changed shapes like our Moon, which supported a Sun-centered universe.

Galileo published his observations in 1610. Many ideas that once seemed certain began to crumble. The universe was suddenly larger, more complex, and far less focused on us. Astronomers refined telescopes continuously. Christiaan Huygens discovered Saturn's rings using improved lenses in 1670. Soon after, Isaac Newton redesigned the instrument using mirrors instead of lenses, allowing for greater clarity and larger construction. Each improvement extended the human gaze further into the unknown.

The Microscope: Revealing a Hidden Universe

While telescopes opened the heavens, another transformation was happening on a microscopic scale. In the late 1500s, Dutch craftsman Zacharias Janssen built one of the first microscopes. It was a curious device that magnified small objects into something more visible. At first, people did not realize how far this tool could take them.

Then came Antonie van Leeuwenhoek. Without a formal scientific education, he created powerful single-lens microscopes through extraordinary craftsmanship. When he examined a drop of water, he saw tiny moving organisms that no human had ever seen. He called them "animalcules." They were alive, active, everywhere, and completely invisible to ordinary sight.

This discovery reshaped our understanding of life. The world was not a fixed stage of solid objects. It was dynamic and busy, driven by life that existed beyond human perception. Leeuwenhoek's work became the first step toward microbiology and modern medicine. Beneath our normal sense of scale, he found an entire universe.

The Space Age: A New Horizon for Telescopes

Even the best telescopes on Earth are affected by our atmosphere. Air bends light, and stars appear distorted or blurry. To escape these limitations, astronomers needed to rise above the sky itself.

In 1990, the Hubble Space Telescope was launched into orbit. Its large mirror captured sharp images of galaxies, star clusters, and nebulae billions of light-years away. Hubble's photographs did more than support scientific theories. They revealed a universe of breathtaking beauty, helping us imagine the earliest ages of cosmic existence.

In 2021, another step forward arrived. The James Webb Space Telescope was designed to see deeper into space using infrared light. This allows Webb to detect galaxies that formed shortly after the beginning of the universe. Webb may even help us understand the origins of stars, planets, and perhaps the first conditions that allowed life.

Instruments of Curiosity: A Never-Ending Quest

The telescope and microscope share the same birthplace. Human curiosity. One turned our attention upward to the stars. The other looked downward into the building blocks of life. Together, they dismantled the boundaries of human perception.

These tools remind us that the world is never limited to what our eyes can see. Every discovery brings new questions. Every question pushes us to innovate. As long as curiosity exists, humanity will continue to explore both the infinite and the invisible.

Our journey of discovery has no finish line. The universe is waiting in every direction.

Chapter 18
Photography

I once watched a tourist point his phone at the moon.

No telescope. No lens attachment.

He snapped the photo, smiled at the screen, and walked away.

I stood there stunned.

Not by his picture.

But by the audacity of the act.

A human being had just tried to pocket the moon.

For most of human history, even kings could not keep a moment. The face of a child, a battlefield's fear, a wedding's joy, all slipped away as quickly as they appeared. Stories try to preserve them, but stories fade. Memory bends and breaks.

Today, anyone can freeze time.

We can keep a moment, share it, return to it, enlarge it on a wall, or carry it in a pocket.

We can press pause on life itself.

This realization always brings me back to a single question:

Who first thought to trap light and make it stay?

Drawing with Light: A Dream Before Its Time

Photography comes from Greek words: *phos*, meaning light, and *graphein*, meaning to write.

To write with light.

A poetic definition, yet the idea existed long before the name.

Humans have always fought against forgetting.

We carved caves.

We inked scrolls.

We painted portraits.

But in all these, a hand acted as interpreter.

The world passed through a person before it entered memory.

Could reality record itself directly?

Ancient thinkers experimented with camera obscura, a simple darkened box or room where light slipped through a tiny opening and formed an upside-down image of the outside world inside it. Chinese philosophers and Greek scholars documented it. Leonardo da Vinci studied it. Artists traced the projections it offered.

Yet the image vanished when the light disappeared.

A ghost with no memory.

The puzzle remained:

Could light be convinced to stay?

The First Photograph: Sun Writing from a French Window

The early 1800s offered the first breakthrough.

Joseph Nicéphore Niépce, a French inventor, coated a metal plate with bitumen and placed it inside a camera obscura. Through his window in Burgundy, light slowly began to mark the surface.

After eight long hours, a faint but permanent picture emerged.

It was titled View from the Window at Le Gras.

The first photograph in human history.

Niépce called it heliography, meaning sun writing.

The world was no longer only experienced or remembered.

It could now be captured.

The achievement was astonishing.

Yet the process was painfully slow, and Niépce died before the world realized what he had created.

But a spark had already taken hold.

Daguerre's Alchemy: Turning Light into Likeness

Louis Daguerre, a theatrical artist fascinated by illumination, advanced what Niépce began. He used silver-coated copper plates, iodine fumes, and mercury vapors to speed up exposure and reveal incredible detail.

In 1839, the daguerreotype was unveiled to the public.

The French government announced photography as a gift to civilization.

Light had become obedient.

Memory could be mechanical.

The technology spread across continents. Samuel Morse, known for Morse code, introduced it to the United States. Portrait studios appeared everywhere. Families who could never afford paintings finally held images of loved ones in their hands.

Photography moved from alchemy into everyday life.

Time was no longer a stranger that always walked away.

From Chemistry to Culture: Eastman and the Democratization of Memory

Early methods still demanded technical skill and speed. Chemicals dried quickly. Mistakes ruined plates.

George Eastman, a bank clerk with an inventive mind, dreamed of simplification. In 1880, he introduced dry plates. By 1888, he launched a new camera called Kodak, preloaded with film.

His slogan was simple:

You press the button; we do the rest.

Photography became a universal language.

Vacations. Birthdays. Soldiers before departure.

Life itself became documented, not just remembered.

In 1900, the Kodak Brownie made photography affordable for ordinary households. The camera became part of growing up.

Humanity shifted from witnessing history to recording it.

The Digital Shift: Pixels Instead of Paper

In 1975, Steven Sasson, an engineer at Kodak, built something no one understood at first. A camera with no film. A sensor called CCD captured images electronically. One image stored on a cassette tape, barely visible by today's standards.

Resolution: 0.01 megapixels.

Impact: world-changing.

The film era began to decline.

By the late twentieth and early twenty-first century, cameras merged with computers and eventually with phones. No chemicals. No waiting. Just instant images stored in invisible clouds. Artificial intelligence and computational photography expanded what images could do. Filters, enhancements, and manipulations blurred the line between what is real and what we wish were real.

Photography no longer only records truth.

Sometimes it edits it.

What Photographs Really Are

Today, anyone can take a picture of the moon.

Or capture a reflection of the moon in a puddle, in an eye, on a car window.

We zoom into details once invisible.
We watch live footage from crises unfolding far from home.
We witness history at the speed of light.
But one question remains:
What exactly do we capture?
A photograph is more than pixels or chemicals.
It is a memory with a pulse.
A fragment of reality preserved with intention.
A moment stopped by a fingertip.
Photography is both truth and interpretation.
A mirror and a message.
A memory and a possibility.
Humanity did not just learn to freeze time.
We taught light how to remember.

Chapter 19
The Thought of Cinema

Cinema began not with sound or color, nor with grand storytelling. It began with a horse and human curiosity.

On December 28, 1895, people gathered inside the Salon Indien du Grand Café in Paris. The room dimmed. A screen brightened. Then a steam train appeared and moved closer, louder, and larger with each passing moment. Some viewers panicked. Others ducked, convinced the train would burst from the screen.

It was only fifty seconds of moving images. Yet that moment shifted human expression into a new form. The Lumière brothers called their creation The Arrival of a Train at La Ciotat Station. Cinema had just opened its eyes.

To understand how humanity reached that night in Paris, we travel back almost two decades, to a problem no one could solve with the naked eye. How does a horse run

The Question That Created Motion Pictures

In 1878, inventor Eadweard Muybridge accepted a challenge from Leland Stanford, a former governor of California. Stanford claimed that during a gallop, all four hooves of a horse left the ground. No one could prove it because the motion was too fast.

Muybridge placed twenty-four cameras along a racetrack. Each camera is connected to a tripwire. As the horse sprinted past, each shutter fired. When the images were placed in sequence, they moved. And for a brief instant, all four hooves hovered above the ground.

A scientific question unexpectedly unlocked a new visual language. Images could move. They could reveal truths beyond human perception. And soon, they could tell stories.

Around this same time, Thomas Edison pursued a different idea. His Kinetoscope allowed one person to peer through a small window and watch a short film loop.

Businesses opened, coins dropped into machines, and pictures moved. But the experience was solitary. Wonder felt too restricted.

When Moving Images Became a Shared Experience

The Lumière brothers saw what Edison missed. They created the Cinématographe, a single device capable of recording, developing, and projecting film. Most importantly, it brought audiences together.

Crowds gathered. Films spread rapidly across cities and borders. A stage magician named Georges Méliès attended one of the screenings and perceived a new possibility. He realized cinema could stretch reality, not just document it. In 1902, Méliès released A Trip to the Moon, sending viewers into a fantasy where rockets struck the Moon's face. Cinema became more than observation. It became an imagination.

Silent films evolved into expressive performances. Gestures, music, and movement formed emotional storytelling. Charlie Chaplin made the world laugh and ache without speaking. Through him, cinema revealed the simplicity and depth of ordinary human experience.

Then movies gained a voice.

Sound, Color, Spectacle: Cinema Refuses to Sit Still

In 1927, The Jazz Singer premiered with synchronized dialogue. Silent films rapidly declined. Sound reshaped narratives and opened new creative paths.

Color soon followed. Cameras became more agile. Special effects advanced. Storytelling expanded. Directors began reshaping how we perceive reality itself.

Orson Welles explored structure and power in Citizen Kane

Alfred Hitchcock made suspense psychological

Akira Kurosawa blended philosophy with action

Satyajit Ray captured real human dignity

Walt Disney gave drawings emotion and breath

By the late twentieth century, filmmakers like Steven Spielberg and George Lucas turned cinema into an immersive spectacle. James Cameron took viewers to the bottom of the Atlantic and then to new worlds like Pandora. Digital effects transformed fantasy into a visible possibility.

Cinema became not just a medium but a universe.

Cinema Now: The Screen is Everywhere

Today, viewers experience cinema in IMAX halls or from handheld screens at home. Filmmaking has embraced virtual reality, performance-capture, and artificial intelligence. Digital characters mimic real people. Holograms perform to living audiences. The boundary between real and imagined keeps fading.

An industry that started with a single horse now generates more than one hundred billion dollars each year. From Hollywood to Bollywood to Nollywood, cinema influences culture and identity across the planet. In 2022, Avatar: The Way of Water surpassed two point three billion dollars in worldwide earnings. But money is not its true measure.

The power of cinema still lives in a moment of awe. A child staring at a screen, transported into a story.

Cinema has become a shared language. It reflects who we were, who we are, and who we might become. It shows thoughts made visible, stories we do not just watch but experience.

The train still rushes toward us. We no longer scream. We simply lean forward, ready to be part of the story again.

Chapter 20
When the Voice Became Wood, Reed, and Brass

Human thought has always explored possibilities outside the body. After studying the stars and shaping tools, we turned attention inward. One question emerged: How does a simple breath turn into speech or song?

This mystery begins in the mind. The brain sends signals to the lungs. Air rises through the trachea to the larynx, where the vocal cords rest. These cords vibrate when air passes through them, creating a buzzing tone. It is not yet a language or melody. The mouth, tongue, lips, and nasal passages shape that vibration into sounds filled with meaning or emotion.

Through this delicate coordination, a thought becomes a voice. And through voice, humans connect.

Yet the connection was not enough. We wanted to extend our abilities beyond the limits of our bodies. We built instruments to imitate or amplify what the human voice could do.

The Instruments That Echo the Human Voice

Reed instruments such as clarinets, harmoniums, and oboes operate like the vocal system. Air moves through a reed that vibrates, producing sound. The body of the instrument shapes that sound into different pitches, much like the throat shapes speech.

In this design:

The reed acts as the vocal cords

The mouthpiece acts as the throat

The keys act as the tongue guiding pitch

Brass instruments rely even more on the human body. Trumpets, trombones, and tubas use buzzing lips as the sound source. The instrument expands and colors the tone, making human breath powerful enough to fill a concert hall.

The pipe organ represents voice on a massive scale. Each pipe resembles a shaped throat, each producing a unique frequency. With air and engineering combined, the organ becomes an entire choir controlled by a single person.

The twentieth century introduced new ideas. The theremin allowed music without physical touch. Sound shifted based on hand movement and intention, as if melody came directly from the air. Later, synthesizers used electricity and digital signals to simulate or transform the human voice. Music no longer required breath, yet still relied on principles learned from vocal expression.

The Boundless Power of Human Thought

It is remarkable that a question about speaking led to the creation of technologies that fill stadiums and stream into headphones around the world.

Voice, once confined to a body, expanded outward. It became louder than lungs, more enduring than a lifetime. Instruments captured and extended emotion. They built entire musical cultures, from whisper-soft ballads to powerful symphonies.

This journey reveals a simple truth. Human thought does not accept limits. It imagines new possibilities, shapes them into tools, and evolves constantly.

A breath became a note. A note became a choir. A voice became music.

The creativity that began with one small muscle in the throat continues to reach upward, always seeking the next sound.

Chapter 21
The Man Who Dared to See What Others Ignored

Leonardo da Vinci

A bird passes above me, wings beating in a rhythm too precise to be an accident. I pause, and a thought rises quietly: what if that motion could be captured? *What if* the invisible flow of air lifting those feathers could be understood and drawn? This moment has belonged to thinkers before me. Centuries ago, a man watched the same flight and refused to treat it as something ordinary. He saw a question hidden inside the wings. That man was Leonardo da Vinci.

He did not come from royalty or scholarly privilege. His father was a notary. His mother is a peasant girl. No noble surname. Simply Leonardo from a small village called Vinci. Yet in the rugged hills of Tuscany, he trained himself to notice what others missed. Not just to look but to observe. Deeply. Patiently. With the curiosity of someone who believed every detail mattered.

Leonardo watched the world as though it had a pulse. He noticed folds of skin, swirling water, twisting smoke, and he tried to uncover the laws behind their shapes. He filled notebooks with questions and drawings that came not from routine skill but from a mind that refused to accept ignorance as enough.

While most people studied scriptures, he opened bodies to understand bones and muscles. While others saw clouds drifting above, he imagined machines soaring through them. He examined the chambers of the human heart, the mechanics of gears, and the mathematics inside a curl of hair. His work did not separate art and science. To him, they were the same pursuit: understanding reality.

Walter Isaacson describes him well. Leonardo was not only a brilliant painter. He was an engineer, anatomist, architect, stage designer, and visionary thinker all at once. The man who painted the Mona Lisa with a smile that has outlasted empires also sketched helicopters centuries before any engine existed. He designed armored vehicles, bridges, diving equipment, and even an early form of a mechanical knight. A concept that resembles what we would now call a robot. All of this from someone who had no formal scientific education.

His brilliance, however, was not built from genius alone. It came from restless curiosity. He would abandon a commissioned painting because a new question had captured his attention. How does a woodpecker protect its brain while drilling into trees? How does light bend inside the human eye? Why does the sky appear blue? Every question drove him further.

He wrote many of his notes backwards from right to left. Perhaps to keep ideas private. Perhaps because his left hand found it easier. Those pages were not intended for public display. They were laboratories of thought where sketches collided with theories and where wonder guided the pen more than answers did.

Yet even with all he created, much of his work was left unfinished. Many inventions remained sketches. Many masterpieces remained incomplete. To some, this might look like failure. But Leonardo's unfinished pages show something else. He never stopped beginning. Never stopped questioning. Never stopped exploring possibilities.

He lived in an age without computers, modern tools, or scientific institutions. But his mind was its own institution. He believed the pattern in a leaf could reveal rules that shape the entire world. To him, everything connected to everything else. No object was too small to deserve attention.

Leonardo da Vinci stands as a reminder that progress begins with the courage to look closely. To question what others ignore. To think without waiting for permission. Today, five centuries later, we still study his machines, his anatomical drawings, and his paintings. We admire not only what he created but the way he thought. So when I watch a bird glide through the sky, I do more than admire its grace. I remember the man who stared at wings and asked a simple but revolutionary question. How does this work? And with that question, he dared to lift humanity toward the possibility of flight.

Epiphanies from This Wave

PART 1 – Dawn of Thought

— *From the first scratches on stone, symbols learned to speak. In that moment, silence fractured into words and humanity found its voice.*

— *A tale whispered by fire grew legs of shadow and light. Centuries later, those shadows moved across a silver screen, carrying whole worlds with them.*

— *The iridescence of a peacock's feather was not just beauty; it was nature's secret code, a spark that became the science of light itself.*

— *The spider spun a fragile thread, yet its silk defied steel. From its web, we learned that strength wears unexpected skins.*

— *A simple lens meant to help the weary-eyed opened hidden realms, telescopes unveiled the cosmos, microscopes the unseen within a drop of water.*

PART 2

Machines of the Mind

Inventions That Transformed Society

Chapter 1
Transportation – Wheel

The Wheel: A Thought That Set Civilization in Motion

I sit and watch a wheel rotate with quiet precision. It is everywhere around us: under vehicles, inside machines, in clocks marking time. A simple circle, yet it shaped the world. Without it, the modern age could not exist. But how did such a thought first appear in the human mind? When did early people recognize that circular motion could reduce struggle and unlock movement?

Today, we take the wheel for granted. Yet its creation took thousands of years of curiosity, experimentation, and necessity. From local trade to global industry, the wheel helped humans move, build, and connect civilizations.

The First Wheels Were Not for Travel

The wheel did not begin as a tool for transportation. Long before carts, humans observed logs rolling or stones shifting, but they had no practical way to apply that motion. People dragged heavy loads across harsh terrain using raw strength alone.

Then a new idea appeared in ancient Mesopotamia around 3500 BCE: the potter's wheel. A rotating platform made shaping clay far easier. It did not carry people or goods yet, but the concept of smooth rotation was established. Human thought took a single principle and looked for more ways to use it.

The next breakthrough required a second piece of insight: the axle. Around 3000 BCE, early carts with fixed wooden wheels and axles changed trade and farming. What once required days of dragging could now be moved in hours. Cities connected. Goods traveled farther. Exchange expanded.

This was the moment motion turned intelligent.

The Wheel Became a Force of Civilization

Once movement became easier, progress accelerated.

Egyptians redesigned wheels for chariots, changing warfare and political power.

Greeks and Romans improved roads and transport, using wheels to stretch their influence across continents.

As centuries passed, wheels entered machines that powered daily life: pulleys lifted water, gears timed the day, waterwheels drove early factories.

The wheel became more than transport. It became a partner in production.

By the 18th and 19th centuries, the world was transformed again.

Trains raced across countries.

Bicycles increased personal freedom.

Automobiles reshaped cities and lifestyles.

Society reorganized itself around the wheel. Roads, markets, industries, and even urban planning shifted to support movement. Production scaled. Distance shrank. What once separated people now connects them.

The Wheel Reimagined for a New Age

The modern wheel is no longer limited to roads.

It powers turbines in aircraft engines, enabling humans to fly at immense speeds.

In hospitals, precise gear systems allow robotic surgery.

On Mars, rover wheels explore distant terrain in search of answers about life.

The future pushes the wheel even further.

High-speed pods in Hyperloop projects reduce friction to near zero.

Electric vehicles refine balance and efficiency using advanced sensors.

Autonomous cars depend on perfectly engineered wheels guided by artificial intelligence.

Technology changes every year, yet the principle remains unchanged: rotation creates possibility.

A Thought That Still Moves Us

From the first spinning clay in Sumer to rovers gliding across Mars, the wheel has carried human ambition forward. It symbolizes the constant desire to reduce effort and expand potential.

As I watch a car pass by with simple grace, I am reminded that civilization began moving forward because one person once wondered:

What if a circle could take us further?

Chapter 2
The Apple That Launched Us into the Stars

Isaac Newton

Long before rockets pierced the sky and satellites orbited above our heads, a quiet moment under a tree changed humanity. An apple fell. Isaac Newton watched, not just with eyes but with a mind tuned to hidden forces. He did not simply witness gravity. He understood it.

Newton was born on Christmas Day in 1642 in Woolsthorpe, Lincolnshire. He arrived early and was so fragile that people doubted he would survive his first night. His father had passed away months before his birth, and when Newton was only three years old, his mother remarried and left him in the care of his grandparents. These early fractures shaped him. He grew inward, searching for stability within thought rather than in people. He was quiet, solitary, and endlessly curious. His mind sought patterns in everything around him.

At Trinity College, Cambridge, he initially showed little promise. Yet when the plague forced the university to close, Newton returned home. Isolation became a laboratory. During two extraordinary years, often called his "miraculous year," Newton laid the foundations for calculus, experimented with light and optics, and began forming the idea that would redefine our universe. The apple's fall became a doorway into a deeper question. Why down and not sideways? Why did it accelerate? Could the same force be acting far beyond Earth?

In his groundbreaking work, Philosophiæ Naturalis Principia Mathematica, Newton explained the law of universal gravitation. He revealed that the same mathematical rules guiding the apple's descent also governed the orbit of the Moon and the paths of planets. A single set of physical laws applied everywhere in the cosmos. Newton gave humanity a framework so precise that modern space missions still rely on its equations.

He once wrote, *"If I have seen further, it is by standing on the shoulders of giants."* Yet very few stood as high as Newton himself. His laws of motion shaped every rocket launch.

The third law, stating that every action has an equal and opposite reaction, remains the reason a spacecraft, once pushed hard enough, escapes Earth's pull and enters orbit. It stays suspended above us, a human-made satellite circling the world.

Biographer James Gleick described Newton as a man who lived largely inside his mind. He was driven by faith, obsessed with alchemy, fiercely competitive in scientific disputes, and often difficult. He was brilliant and flawed, complex and human. Yet at his core lived an unstoppable curiosity about how the universe works.

We can imagine a different history if he had never looked up after that apple fell. Would humanity have reached space? Would satellites still guide our navigation? Would we have dared to aim for the Moon? Our age of space is indebted to a man who never traveled far from his homeland but who gave us the mathematics to leave the Earth behind. When satellites glimmer in the night sky, or when rockets rise with fire and ambition, Newton's voice is still present. His ideas continue to propel us upward. Centuries later, his equations remain alive in every orbit, every launch, every dream pointed toward the stars.

Chapter 3
Transportation – The Thought That Moved the World

A distant train whistles into the evening, the sound slowly folding into the wind. There is something about that call of movement that feels deeply human. A symbol of progress. A reminder that the world is always shifting. Yet long before trains carved paths across continents, the idea began quietly as a question in the human mind: Can we move the world with steam?

Today, we watch high-speed trains glide with precision. But this achievement did not begin with steel rails or polished engines. It began with fire, water, and imagination.

From Fire and Water to Motion

Before highways and aircraft, humanity needed to solve a problem. Walking and using animals for transport were not enough for growing societies. People needed to move faster, travel farther, and carry more without exhausting human or animal strength.

In the 1700s, inventors began exploring how heat could be turned into useful energy. Early steam engines were first used to pump water from mines. Yet within those engines lived a new possibility. Someone eventually asked a transformative question: Could this power move more than water?

James Watt refined the early steam engine into a reliable source of motion. His work marked a turning point in history. For the first time, humans created machines that did not rely on rivers, wind, or muscle.

They had captured energy inside a system that could run anywhere steam could be produced.

The Leap to Locomotion

The next step was bold but logical. If steam could move pistons and gears, why not people and goods across land? In the early 1800s, inventors like Richard Trevithick and George Stephenson pursued this vision. Their first locomotives were noisy and imperfect, but they worked.

When Stephenson's engine, called the Rocket, succeeded at the Rainhill Trials in 1829, the world witnessed a new future. Travel that once required days or weeks on horseback could now be measured in hours. Railroads began spreading outward in every direction.

Commerce accelerated. Cities expanded. Human speed increased. The rhythm of life changed.

Steam did not simply move objects. It pushed society forward.

A New Shape of the World

Railways reshaped global relationships. Tracks connected distant regions. Economies that once operated separately became linked. New towns formed near stations and along rail lines. The physical size of nations seemed to shrink. Distance was no longer an obstacle. It was a path.

Over time, technology has evolved. Steam engines gave way to electric and diesel trains. The twenty-first century introduced magnetic levitation and bullet trains that glide almost silently across the land. Yet the original idea remains present inside every modern carriage. Energy turned into motion. Motion turned into possibility.

Where We Might Go Next

Trains today are precise, fast, and increasingly sustainable. Engineers experiment with renewable power sources and automated systems. Some imagine networks that could circle continents in hours. The thought that began in a mine, pumping water in the dark, now reaches every corner of our global society.

A simple question changed everything. How can we move the world better?

Some thoughts, once set in motion, continue to travel forward without end.

Chapter 4
Electricity – The Thought That Conquered Darkness

There was a time when nightfall meant stillness.

When darkness spread, humanity surrendered and waited for dawn.

Then a question emerged inside the human mind.

Can light be captured and commanded?

Electricity became the thought that reshaped civilization.

Not only wires and sparks, but belief that the unknown could be understood.

From Franklin's lightning experiment to Tesla's power networks, that thought transformed the world.

How Humans Discovered the Sky Held Power

The 18th century produced Benjamin Franklin, a curious thinker who suspected that lightning and electricity were linked.

In 1752, he conducted his famous kite experiment. A metal key attached to a wet string attracted a spark from the storm above.

The sky was no longer just noise and fear.

It was energy that could be studied.

Franklin helped define charges, conductors, and the hidden force inside storms.

It was the first step in turning a mystery into a science.

The Early Race to Create Light

Firelight had guided humans for millennia.

Candles and gas lamps could break darkness, but only briefly.

A cleaner and steady light became a growing desire.

In 1802, Humphry Davy created the first electric glow.

Later he produced arc lighting, bright enough to startle crowds but too harsh for homes.

Progress continued across nations.

Inventors shaped metal, glass and wires.

They failed often, learning a little more with every mistake.

Edison's Search for the Right Filament

Thomas Edison believed the secret to a lasting bulb was inside the filament, not the electricity.

His team tested material after material until bamboo charcoal produced a warm, durable glow lasting more than one thousand hours.

In 1879, a reliable electric light was achieved.

It proved that patience combined with thought can conquer darkness.

The Power Delivery Battle, DC and AC

Once bulbs existed, electricity needed to travel.

Edison supported direct current (DC), which could only move short distances.

Cities needed power lines everywhere to keep the current strong.

Nikola Tesla offered alternating current (AC), which traveled farther with less loss.

Supported by George Westinghouse, his system eventually powered the Chicago World's Fair in 1893 and the Niagara Falls power plant in 1896.

AC systems spread across continents.

Cities brightened.

Factories extended work hours.

Homes became safer at night.

The modern electrical network was born.

Electricity Becomes a Foundation for Civilization

Grids grew.

Electricity reached schools, hospitals, and farms.

It supported science and industry.

It powered dreams.

Today, electricity is everywhere.

It charges communication, protects health, and energizes economies.

It has become the lifeline of modern life.

A Thought That Continues to Evolve

Electricity began as a curious spark.

Humanity questioned the dark and refused to accept limits.

Every glowing bulb reminds us of this truth: the smallest thought can change the entire world.

Chapter 5
The Relentless Spark That Lit the World

Thomas Edison

A silent workshop.

A single bulb unfinished.

A man refusing to stop.

Thomas Edison did not wait for inspiration.

He created it through effort.

He struggled through school and left with little praise.

Yet he learned by touching, tinkering and daring to experiment.

Telegraph machines became his teachers.

Failure became his method.

His Menlo Park laboratory was not clean or peaceful.

It was filled with broken glass, burnt wires, and tired hands.

More than one thousand attempts were made before the right bulb appeared.

He did not complain.

He studied every failure as progress.

His belief was simple.

Success belongs to those who continue.

He called genius one percent inspiration and ninety-nine percent perspiration.

He lived by that rule with long days and constant note-taking.

Menlo Park became the first industrial research center.

Ideas were tested.

Tools were shared.

Innovation became a system.

There Edison advanced:

- The phonograph
- Motion picture cameras
- The carbon microphone and other communication tools

Biographer Edmund Morris described Edison as a paradox.
Nearly deaf, yet the inventor of sound recording.
Not elegant in speech, yet powerful in ideas.
He did not drift like dreamers.
He built solutions that worked.
Every modern room with light carries his story.
Human effort turning darkness into clarity.
His life reminds us that invention is not luck.
It is a vision held tightly through every failure.
Edison proved that determination itself is a spark, and that spark can light the entire world.

Chapter 6
Nikola Tesla – The Man Who Listened to the Future

Nikola Tesla was more than an inventor.

He was a thinker who saw electricity not as a tool but as a language spoken by the universe.

Born in 1856 in a small village in present-day Croatia, Tesla's childhood imagination was already alive with energy. Long before the world understood alternating currents or wireless signals, he saw visions of invention that arrived in flashes of clarity. These ideas appeared so complete in his mind that he could design machines without ever drawing them on paper.

Tesla travelled to America in 1884, carrying only a letter of introduction and a brilliance few could comprehend. For a short time, he worked with Thomas Edison. Their minds clashed. Edison trusted what he could hold and test one failure at a time. Tesla believed in what he could see inside his mind with perfection and precision.

At the heart of their disagreement was electricity. Edison defended direct current, DC, while Tesla championed alternating current, AC. The debate was more than scientific. It was a decision about how the world would be powered. Tesla's AC could travel long distances and support expanding cities. Eventually, AC became the global standard, lighting streets and shaping industry.

But Tesla's curiosity did not stop with wires or generators.

He imagined a world where energy flowed through the air.

He dreamed of transmitting power without cables.

He believed information could travel across oceans in moments.
Many of these ideas were too advanced for commercial investors of his time.
Today, wireless technology, remote control systems, radio communication and radar all reflect foundations that Tesla quietly laid.
His life was filled with achievements but also sacrifices.
He often struggled financially.
Ideas were taken, credit was given to others and the world did not always recognize what he had offered.
In later years, Tesla lived mostly in hotels in New York City.
He spent his days feeding pigeons outside and writing notes about inventions he still hoped to finish.
He spoke several languages.
He slept very little.
He cared deeply for the natural world and believed knowledge was a responsibility, not a possession.
Nikola Tesla died in 1943, alone and overlooked.
Yet today, his name powers an era. His patents support the technologies beneath modern life.
Electric vehicles carry his name forward.
Wireless signals circle the globe because his vision crossed the limits of imagination.
Tesla walked ahead of his time.
He listened to the future as if it were already speaking.
And even now, the world continues to catch up to him.

Chapter 7
The Thought of Telephones: Voices Across Wires

Once, a voice could travel only as far as the ears around it.

Walls separated people.

Distance turned conversations into silence.

Messages crawled on paper for weeks before they reached their destination.

In that silence, a thought began to form.

Could the human voice leap beyond distance?

Could sound travel through wire and arrive in another room, another town, another country?

Alexander Graham Bell found himself at the center of that question.

Born in 1847 in Edinburgh, Scotland, Bell grew up surrounded by sound. His father taught the science of speech. His mother, who was deaf, taught him something deeper. Communication is not only sound. It is connection, attention and care.

This shaped Bell's path.

He studied the vibrations of voice.

He helped the deaf communicate.

And quietly, he worked on an idea that promised to change the world.

By the 1870s, Bell was in Boston, teaching and researching. The telegraph already existed, but could only send coded messages. Bell imagined something alive and human. He believed that sound could be turned into electrical signals and then rebuilt as speech on the other side of a wire.

In 1876, that belief became reality.

His first words through the telephone were simple.

Mr. Watson, come here, I want to see you.

His assistant heard his voice, clear and immediate.

A human voice had travelled farther than ever before.

The invention changed communication forever.

Not through symbols or dots.

Through speech itself.

People could speak across distances in real time.

Families stayed connected.

Businesses accelerated.

Nations negotiated directly instead of waiting for letters to arrive.

The first telephone line linked Boston and Somerville.

Soon networks spread to cities everywhere.

At first a luxury, the telephone slowly became a part of everyday life.

Distance began to feel smaller.

Voices were always within reach.

Technology advanced.

Rotary phones gave way to digital systems.

Wires reached across continents and under oceans.

Eventually, voice travelled without wires, through radio waves, satellites and the internet.

The smartphone, a device that holds communication, information and memory all in one hand, began with Bell's original insight.

Sound can travel.

Thought can move.

Connection can overcome separation.
The telephone is more than a machine.
It is a bridge between people.
A reminder that the human voice matters and deserves to be heard.
It began as a thought.
A voice calling out.
A wire carrying hope across the distance.

Chapter 8
The Thought of Automobiles – Personal Freedom

Karl Benz's early automobile

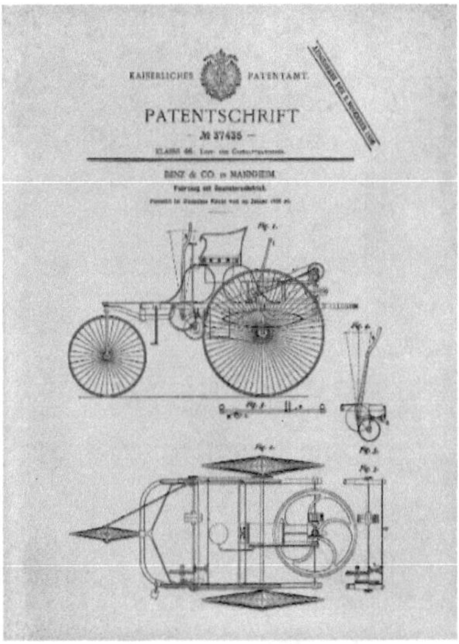

Patent drawing of the Benz Patent-Motorwagen, (1886)

At first glance, it looked like a sewing machine with wheels.

No steering wheel. No roof. No windshield.

Three wire spoked wheels and a primitive internal combustion engine.

Yet inside that humble form was a thought powerful enough to change how humanity moved.

Karl Benz's Patent Motorwagen of 1885 marked the beginning of a new direction.

But inventions often need courage to truly come alive.

That moment came from someone who believed before the world believed.

His wife.

Bertha Benz and the First Journey

One morning in 1888, Bertha Benz quietly rolled the Motorwagen out of their home in Mannheim.

With her two sons beside her, she drove toward Pforzheim, her mother's hometown.

The journey was more than 100 kilometres long.

There were no maps for cars, no petrol stations, no mechanics waiting to assist.

She bought fuel from a pharmacy.

She used a hat pin to open a clogged fuel line.

A cobbler repaired the worn brake pads.

This became the first long-distance drive in history.

The first road test.

The first demonstration.

The first advertisement.

And it worked.

People watched in astonishment as a carriage moved without horses.

The idea suddenly became real.

The Promise of a Machine That Moves Itself

The automobile represented more than transportation.

It represented independence.

People could travel between towns without relying on animals or fixed tracks.

It stretched the human footprint beyond the limits of walking distance or public schedules.

Years later, Henry Ford changed not what a car was, but who could own one.

Henry Ford and the Global Road

The Model T in 1908 and the moving assembly line in 1913 transformed the automobile into a product for ordinary people.

Cars stopped being rare machines for the elite.

They became part of everyday life.

Freedom expanded:

- Families visited distant relatives
- Workers lived farther from factories
- Roads connected places once isolated

New industries appeared.

Fuel stations rose along highways.

Rubber plantations grew.

Cities spread outwards and suburbs were born.

Hotels evolved into motels.

Homes welcomed garages.

Drive through restaurants emerged.

A car was no longer an invention.

It was a lifestyle.

The Cost of Progress

With millions of cars came consequences.

Cities choked with traffic.

Skies filled with smog.

Oil shaped global politics, sometimes through conflict.

The machine that once symbolized escape now contributed to urban stress and environmental strain.

Humanity did not step back. We adapted.

Electric vehicles. Hybrid engines.

Today, artificial intelligence is even learning to drive us.

A Thought That Keeps Moving

The automobile remains a symbol of movement and autonomy.

It began as a strange tricycle.

It became a global force that redesigned how we live, work and explore.

It is a reminder that sometimes the simplest way to move forward can carry the world with it.

Chapter 9
Symbols: The Silent Messengers of Our Thoughts

One day, I sat behind the wheel waiting at a traffic light.

My eyes wandered to the chrome shapes attached to the backs of cars.

Circles. Stars. Letters. Animals.

They appeared silent, but they were speaking.

These were not decorations.

They were ideas transformed into shapes.

Thoughts that survived through centuries.

Humans have always searched for ways to express and preserve meaning.

Long before alphabets, we painted symbols on cave walls.

The First Language of Meaning

In Lascaux, France, deep underground, where sunlight never reached, early humans drew horses and bulls.

These paintings were not idle sketches.

They captured emotion, belief, and hope.

A message to those who would come later.

From those first marks, systems of writing evolved:

- Hieroglyphs in ancient Egypt
- Cuneiform in Mesopotamia
- Pictograms in early China

Every character was a fingerprint of human thought.

When Signs Became Sacred

As religions formed, symbols gained power.

A cross represented salvation.

A crescent reflected divine guidance.

The Om carried the vibration of the universe.

Symbols no longer needed translation.

They spoke directly to the soul.

But power can corrupt meaning.

The swastika, once a sign of prosperity in India, was twisted into a symbol of hatred by the Nazis.

Even the purest ideas can be stolen and reshaped by ideology.

A New Era: Brands as Belief Systems

The 20th century built a new temple: the marketplace.

Now companies needed symbols.

Coca-Cola's script carried comfort and nostalgia.

Nike's swoosh suggested motion and daring.

Apple's bitten fruit invited curiosity and rebellion.

We do not wear brands.

We wear identity.

We do not buy products.

We buy what they represent.

Cars turned roads into galleries of philosophy:

- The four rings of Audi tell the story of unity between the four companies
- Hyundai's H represents a handshake of trust
- Mercedes's star points toward ambition on land, sea and air

- BMW's circle reflects the Bavarian flag and its engineering pride
- Toyota's three ellipses connect the customer, company and the world
- Ferrari's stallion honours bravery from a wartime pilot
- Lamborghini's bull reflects untamed strength and confidence
- Rolls-Royce offers the Spirit of Ecstasy as a movement with elegance
- Tesla's T reveals a motor's internal cross-section, a tribute to engineering
- Volkswagen remains the simple people's car, timeless in its clarity
- Ford carries a signature that became global reliability

Every shape carries a story.
Every symbol carries a decision.

Symbols That Move Societies

Movements discovered their own shapes:

- The peace sign
- The rainbow flag
- The raised fist

These symbols did not follow trends.
They created change.

A New Language in the Digital Age

Today, our everyday symbols are pixels.

Emojis deliver joy and sorrow across continents instantly.

Apps are identified by small icons that serve as digital doorways:

Instagram's camera.

Facebook's single letter.

Twitter's bird in flight.

What comes next

Perhaps holograms.

Shapes that respond to our voice.

Symbols that activate with a glance.

But one truth remains constant.

A symbol is a memory that has learned to speak without words.

It is a universe compressed into a mark.

We do not just live surrounded by symbols.

We live through them.

Because humans always hope to be seen, understood and remembered.

Chapter 10
They Made a Toy Given by Their Father a Reality for the World

A thin white line stretched across the morning sky, traced by a distant jet. I paused and looked up. For a moment, the sky did not feel unreachable. Humans have looked upward for centuries, asking one question. How did we learn to fly? For two brothers, the journey began with a small toy helicopter powered by a rubber band. A simple gift from their father. Most children would smile and forget it. Wilbur and Orville Wright did not forget. They carried that moment into history.

The Thought of Flight

Human fascination with flight is ancient. Stories like Icarus imagined wings before technology could create them. Leonardo da Vinci sketched flying machines that studied lift and balance long before they were possible. He explored the mechanics of air, motion, and control. His concepts waited centuries for the right minds to arrive.

Imagination alone could not conquer the sky. Flight demanded answers to three challenges. Lift. Propulsion. Control. Each needed both creativity and science.

Bicycle Mechanics Who Outthought the World

Wilbur and Orville Wright did not attend elite universities. They did not have armies of assistants. They repaired bicycles in Dayton, Ohio. Yet their thoughts were sharper than the metal they repaired. They studied bird wings, tested designs for stability, and built a wind tunnel with their own hands.

Many tried to fly by building larger gliders. The Wright brothers focused on control. They realized that a pilot must guide the aircraft like a bird adjusts its wings. They created a system that allowed the flyer to twist and maneuver in the air.

On December 17, 1903, at Kitty Hawk, North Carolina, their plane lifted from the ground. It stayed in the air for twelve seconds and traveled 36.5 meters. History had taken flight.

Beyond That First Lift

At first, the world doubted. Even after the flight was proven. But the brothers continued improving their designs. By 1905, the Flyer III stayed airborne for more than thirty minutes and traveled nearly thirty-nine kilometers. Flight had evolved from a fragile experiment into transportation.

Others soon joined the journey. Glenn Curtiss produced stronger engines. Louis Blériot crossed the English Channel in 1909. Alberto Santos Dumont inspired crowds in Europe and Brazil. Airplanes became a pathway into the future.

Aviation Becomes a Necessity

The first scheduled commercial flight took place in 1914 from St. Petersburg to Tampa in Florida. Short distance, great impact. Wars accelerated innovation. World War I introduced aircraft in combat. World War Two created the jet engine and the possibility of surpassing the speed of sound.

After the wars, commercial aviation expanded rapidly. Aircraft manufacturers like Boeing and Douglas built planes capable of circling the globe. Flight reshaped business, culture, and human connection.

A Sky That Keeps Changing

The pace of innovation continues. The Concorde explored supersonic travel. Today, drone technology, electric propulsion, and vertical takeoff aircraft are redefining aviation. Scientists explore hydrogen and solar-powered planes for a sustainable future.

The sky remains open for new ideas.

The Legacy That Began With a Toy

This story did not begin in a laboratory. It began with a curious gift. Two boys looked upward and never stopped asking questions. Their thoughts lifted humanity off the ground.

Somewhere today, a child looks at an airplane and wonders what it feels like to fly. From that wonder, a new idea will rise.

Chapter 11
Ocean Giants: Building a Ship That Floats, Breathes, and Moves Cities

A modern ship does not crawl or fly. Yet it carries weight that could crush mountains, floating on water as if gravity changed its rules. A ship is not simply a boat. It is a moving city of steel. It can carry twenty thousand containers, thousands of passengers, or an entire air force.

It begins with one idea. Build something enormous that refuses to sink. Not as a metaphor. As a measurable goal.

From Concept to Construction

Steel is heavier than water. Yet a ship floats because its shape and volume displace water with precision. Engineers plan everything using advanced design software that simulates movement, balance, fuel flow, stability, and the forces of waves. The purpose of the ship defines its form. Cargo. Transport. Warfare. Leisure.

Once defined, construction begins.

Massive steel plates are bent into curved hull sections. Each piece becomes part of a watertight structure. Most shipyards now build in modules, like puzzle blocks. These blocks are joined into a full vessel in yards across South Korea, China, and Germany.

Breathing With the Sea

Balance is continuous work. Ballast tanks inside the ship fill and release seawater to maintain stability. They respond to shifting cargo, rough tides, and powerful storms. The system acts like lungs that adjust to keep the vessel upright.

Deep inside lies the heart. A marine engine that can be larger than a multi-story building. Diesel units, gas turbines, or nuclear reactors drive long propeller shafts that rotate for thousands of hours without failure.

A City on Water

Above the hull, there is architecture. Rooms for passengers. Bridges for navigation. Communication systems. Elevators. Waste treatment facilities. Swimming pools and gardens on cruise ships. Everything must function safely while the sea constantly changes below.

Then there is the bridge, the mind of the ship. Here, the crew uses radar, sonar, and satellite systems to guide the ship with accuracy. Even in the middle of the ocean, a ship knows its position within a meter. Safety tests are extensive. Fire drills, engine stress checks, evacuation systems, and hull protections against marine growth are all required before the ship is released.

When Steel Touches Water

Finally, the ship meets the ocean. It glides into the sea as if it was always meant to be there. No panic. No sinking. Just confidence built on careful thought and precise engineering.

A ship is heavy in material but powered by something light. Human thought. Coordinated effort. Courage combined with calculation. Respect for the sea expressed through design.

We build vessels that seem too large to float. Yet they do. Because thought made them real.

Chapter 12
Traffic Control Systems

A little more than a century ago, the streets of cities were quiet. Horses trotted gently through narrow paths, and pedestrians walked without fear of rushing engines. The late 1800s held a rhythm unhurried by machines. A thriving city full of honking vehicles was still beyond imagination.

Even further back in history, oceans belonged to the winds and waves. Sailors depended on stars, tides, and luck. There were no radar screens or radio towers guiding ships through fog or storms.

Then the 20th century arrived and changed everything.

The skies opened to flight after the Wright brothers lifted a machine into the air in 1903. Roads became crowded with motor vehicles. Oceans witnessed steel ships powered by engines instead of sails.

Movement became faster, broader, and constant.

Yet speed brought a new question.

How do we prevent chaos?

Humanity needed control systems designed to keep travel safe on land, sea, and air. These networks were built gradually, shaped by technology and necessity. Without fanfare, they organize the flow of millions of journeys every day.

In this chapter, we explore how people learned to guide motion across the earth, waters, and skies. From the first city traffic signals to the advanced radar and navigation systems of today, this is the story of how we mastered movement, one rule and one signal at a time.

Chapter 13
The Evolution of Road Traffic Control: From Whistles to Artificial Intelligence

Stand at the corner of a busy intersection. Cars move like synchronized waves. Pedestrians wait, then cross with confidence. Signals blink, cameras watch, and signs direct with silent authority.

This balance was not always here.

Once, roads were dirt paths with no limits, no lanes, and no safety rules. Humans negotiated space with instinct alone. Ancient Roman roads connected an empire but lacked dedicated signals. By the 1800s, carriages filled city streets, and police officers became the first traffic controllers, blowing whistles and waving arms.

Progress began with a simple idea.

Light could guide motion.

The First Lights

In 1868, outside the British Houses of Parliament, J. P. Knight installed a gas-lit traffic signal inspired by railway systems. Semaphores signaled by day and colored lamps by night. The design worked briefly before a gas explosion ended the experiment.

Still, the seed was planted.

True change arrived in the United States.

In 1912, Lester Wire, a police officer in Utah, installed one of the earliest electric traffic lights showing red and green. Two years later, Cleveland introduced the first full traffic light system.

But drivers needed a warning between stopping and going.

Detroit answered that need.

In 1920, William Potts added the yellow light.

Red, yellow, and green became the universal language of the road.

Early lights had mechanical timers. They followed a routine, no matter how empty or crowded the streets were. Traffic obeyed, but efficiency was missing.

Then came smarter ideas.

Coordination and Control

By the mid-1900s, cities experimented with systems that linked multiple signals. Toronto and Wolverhampton led with coordinated timing. Instead of treating intersections separately, movement could be managed on a larger scale.

The 1970s brought sensors into roads.

Inductive loops hidden beneath the asphalt detected vehicles. If no one waited, the light stayed green. Signals gained limited awareness.

Cameras arrived in the 1980s, giving control centers a live view. Operators adjusted traffic flow during surges or emergencies. Roads were no longer blind.

The Age of Algorithms

By the 2000s, artificial intelligence entered the scene.

Los Angeles developed systems that analyzed congestion in real time and adapted signals automatically. London's SCOOT system adjusted timing continuously based on traffic flow. Traffic no longer reacted only to the present. It was prepared for what was coming.

Today, advanced systems recognize:

- Pedestrians
- Cyclists
- Buses
- Emergency vehicles

Signals adjust according to priority and safety.

Cities like Singapore are testing lanes that change direction based on demand, guided by real-time predictions.

A Glimpse of Tomorrow

Autonomous vehicles are emerging. With them comes a possibility no one in 1868 could imagine.

A world without traffic lights.

Cars may communicate directly with one another and with road systems to decide movement. No waiting at red lights. No honking. A continuous flow guided by information instead of physical signals.

From a single gas lamp in London to networks guided by artificial intelligence, road traffic control has become one of the quiet miracles of modern life. It protects us every time we travel, even when we never stop to notice.

Chapter 14
The Evolution of Rail Traffic Control: From Hand Signals to Autonomous Railways

There was a time when a steam locomotive racing along iron tracks stirred both excitement and fear. Early railways had almost no safety structure. No electric signals. No guard rails. Men stood by the tracks waving flags and blowing whistles, trusting that another train was not coming from the opposite direction.

Rail traffic control began as a simple idea. Trains shared a single track, so drivers relied entirely on visual signs from station masters and basic time schedules. It was a fragile arrangement. When one train ran late, the entire system collapsed into chaos, often ending in tragedy.

The Birth of the Block System

As rail networks expanded in the early 1800s, it became clear that trust was not enough. Operators created the time interval system. If one train departed, another waited several minutes before following. This slowed traffic and still failed during unexpected delays.

By the 1840s, rail lines in Britain introduced the block system. Tracks were divided into isolated sections called blocks. Only one train was allowed in each block at a time. Signalmen communicated with flags and semaphores to show whether a block was free.

Then came the electric telegraph. Station operators could send and receive messages instantly. For the first time, a train's location was known beyond eyesight.

New Complexity Underground

In 1863, London opened the Metropolitan Railway, the first underground railway in the world. The tunnels required more advanced signaling because drivers could not depend on natural visibility.

By the 1870s, mechanical interlocking appeared. Tracks and signals were designed so that conflicting instructions became physically impossible. A signal could not turn green if the track ahead was unsafe. Rail control was gaining intelligence.

Central Control and Electrification

The twentieth century reshaped railways once again. Electrification improved performance, and centralized traffic control emerged. In the 1920s and 1930s, control rooms appeared across Europe and the United States. Teams monitored large areas through electric signals and telephone communication. Trains were now guided by a broader view, not just local judgment.

The Age of Automation

Computers arrived in the 1960s. Automatic train control systems measured speed, distance, and braking. Japan's Shinkansen high-speed line was among the first to use fully integrated control, merging safety and speed into a single design.

Real-time digital monitoring was developed in the 1980s and 1990s. Software coordinated thousands of signals and switches, preventing delays from spreading and adjusting routes dynamically.

Automation advanced further. Cities such as Copenhagen, Singapore, and Dubai now operate driverless trains. They stop with higher accuracy, react faster, and reduce energy consumption by optimizing movement.

High-speed systems like China's CRH and Europe's TGV operate within the European Rail Traffic Management System. This standard allows trains from different countries and manufacturers to move across borders under one digital communication language.

What Comes Next

Artificial intelligence is becoming a quiet partner in the control room. Systems predict equipment failures, schedule repairs, and even reroute traffic before problems emerge.

From whistles in the fog to trains guided by satellites, rail traffic control has evolved from human reflex to machine precision. A technology that began as simple caution has grown into a network of prediction and orchestration.

Chapter 15
The Evolution of Sea Traffic Control: From Lighthouses to Global Maritime Intelligence

The sea once appeared infinite. Its rules belonged only to nature. Ancient sailors navigated by stars, winds, and intuition shaped by years of experience. There were no controllers or communication systems. Collisions were avoided not through planning but through the ocean's vast emptiness.

As exploration increased and ships grew larger, oceans transformed from empty space into crowded highways. Trade routes carried wealth, power, and competition. Safety demanded structure.

Lighthouses and the First Signals

Lighthouses became the earliest form of sea traffic control. They warned sailors of coastline hazards and guided them home.

The Pharos of Alexandria, one of the Seven Wonders of the Ancient World, stood as both a light and a message. There is danger here. Navigate carefully.

By the 1800s, ports like London, Rotterdam, and New York were busier than ever. Harbor masters began directing ship movements. Rules of the right-of-way formed. Nautical flags gave ships the ability to communicate across long distances using colors and patterns. The sea gained its first regulated language.

Technology Transforms the Sea

The twentieth century introduced wartime technologies into civilian navigation. Radar allowed ships to detect each other in darkness or fog. Radio communication created real-time contact between vessels and coastal authorities.

In 1948, nations formed the International Maritime Organization. For the first time, global shipping followed the same safety rules. Sea travel began to operate as a coordinated international system.

Data Becomes the Ocean's New Map

In the early 2000s, the Automatic Identification System changed everything again. Every major vessel transmitted its name, direction, speed, and destination. Ships no longer traveled blind. Authorities could see the entire ocean as a living map filled with moving data.

Today, ports such as Singapore, Shanghai, and Rotterdam operate advanced control centers. Artificial intelligence manages vessel movement, monitors weather changes, and prevents congestion before it begins. Satellites cover every continent. Port operations are simulated digitally to test outcomes before ships even arrive.

Drones support surveillance and logistics. Ports have become ecosystems of sensors, networks, and intelligence.

Steering Toward the Future

Autonomous ships are being tested across open waters. Machine learning captains make decisions faster than humans, learning from millions of recorded voyages. New digital sea corridors guide ships along optimized paths.

Sea traffic control has become more than safety. It is efficient, clear, and sustainable.

The waters that once felt unpredictable now follow structures designed by global cooperation.

What began with stars and stone towers now lives in algorithms and satellites. Humanity's knowledge has reshaped the ocean into a coordinated world route, where ships move with confidence from one horizon to the next.

Chapter 16
The Evolution of Air Traffic Control: From Ground Flags to AI in the Sky

Look up. A jet draws a white line across the blue sky, steady and precise. Today, it seems effortless. Yet the history of air travel began with pilots relying on little more than eyesight and courage. No guiding voices, no radar, no digital maps. The sky was once a place of instinct and hope.

The Era of Visual Navigation

In the earliest years of aviation, pilots navigated using the ground beneath them. Landmarks, compasses, and good weather were essential. Communication was minimal, often nonexistent. As commercial aviation grew in the 1920s, aircraft increased in number, and danger followed.

Ground crews tried to impose order. Flags signaled approval for landing or takeoff. Flares were fired to warn of weather troubles. But the skies were becoming busy, and simple signs could no longer keep aircraft apart.

Aviation Takes Structure

In 1926, the United States established designated airways. Pilots and ground stations began using radio to share weather and instructions. A new era began. The sky gained rules.

During the 1930s, the Air Traffic Control Division was formed. Controllers worked from airport towers, tracking movements with paper strips and sharp attention. Aircraft were still monitored by time and expected location, not real evidence of where they truly were.

War Forces Innovation

World War Two introduced radar, created first to detect enemy aircraft. Soon, radar became essential for civilian flight. Controllers could watch aircraft appear as moving points on screens, even when clouds hid the sky. After the war, radar centers spread worldwide. Air traffic control evolved into a detailed, three-dimensional choreography.

Computers Enter the Sky

By the 1960s, computers changed aviation again. Flight plans were entered digitally instead of being handwritten. Radar images were displayed with more clarity and accuracy. Systems like CARDS helped controllers manage larger airspaces with greater safety.

The Aircraft Learns to Speak

The 1990s brought ADS-B, a technology allowing aircraft to broadcast their own position, speed, and altitude. Aircraft became active voices in the conversation, alerting both ground stations and other aircraft. Combined with satellite navigation, this formed a powerful information network. Every plane knew where it was and what moved nearby.

AI Joins the Control Room

In the 2010s, predictive algorithms began assisting controllers. Software could forecast traffic buildups, avoid storms early, and suggest fuel-saving routes with careful precision. Decisions once reactive became proactive.

Today, air traffic control is a partnership between skilled humans and intelligent machines. Controllers manage digital skies with vast data: weather, emergencies, route conflicts, and aircraft performance. Our airways are safer, faster, and more efficient than at any time in history.

The Skies of Tomorrow

We now stand at another turning point. Autonomous flight develops rapidly. Drones deliver packages. Flying taxis emerge in test cities. Researchers imagine free flight systems where aircraft communicate and negotiate space with AI guidance, removing rigid corridors entirely.

From waving flags to lines of code, air traffic control has transformed improvisation into orchestration. What began as a struggle to prevent collisions is now a vision shaping the future of our skies.

Humanity learned not only how to fly, but how to fly safely together.

Chapter 17
The Thought of Space Exploration – Beyond the Earth's Boundaries

From the first night humans gazed upward, the stars have called our curiosity. Ancient astronomers mapped constellations, imagining gods and stories among the specks of light. Their questions hinted at a deeper desire. What lies beyond our world?

For centuries, the sky seemed unreachable, reserved for the divine. Then scientists like Copernicus and Galileo challenged that belief. Earth, they said, is not the center. We are in a tiny world within a vast universe. With that idea, the cosmos became a possible destination.

The Space Age Begins

In the mid-twentieth century, competition pushed humanity upward. The Cold War fueled a race for the stars. In 1957, Sputnik 1 became the first human-made satellite in orbit. It altered history. Space was no longer a fantasy. It was a real frontier.

In 1961, Yuri Gagarin orbited Earth and returned safely. He became the first human to leave our planet and come back with a story, a reminder that limits are often temporary.

Then came a moment the entire world still remembers. In 1969, Neil Armstrong stepped onto the Moon. His words echoed not just across the Moon's surface but into the future. One small step for a man. One giant leap for humanity. The Moon was no longer a distant light. It held a human footprint.

Exploring Without Humans

Not every explorer wears a space suit. Robotic missions like Voyager traveled beyond the planets. Mars rovers examined red soil and ancient riverbeds. Space telescopes revealed galaxies far beyond human reach. Machines extend our senses into places where human life cannot survive.

A Home in the Sky

Today, the International Space Station orbits above us. It is a laboratory and a symbol. Nations cooperate there, studying how humans might live and thrive beyond Earth. Each experiment prepares us for deeper exploration.

Space for Everyone

The idea of leaving Earth is no longer limited to astronauts. Newly developed systems are making spaceflight more accessible. Private companies have begun launching civilians into suborbital space. What was once unimaginable now seems within reach.

The Dream of New Worlds

Mars has become the next great challenge. Some believe that settling another planet might protect our future, giving humanity a second home. The journey will be long, uncertain, and difficult. Yet throughout history, humans have walked toward the unknown rather than away from it.

Are We Alone

The search for life beyond Earth continues. Telescopes examine distant planets for signs of atmosphere, oceans, and biology. Each discovery brings us closer to answering one of our oldest questions.

Space exploration is the story of our relentless curiosity. We are a species that refuses to remain still. We push boundaries. We cross oceans, split atoms, rise into the sky, and now reach for the stars. The night sky still asks us the same question it has whispered since the beginning.

Will you keep going?

Chapter 18
Satellites – Reaching for the Stars

I look up at the night sky, silent, infinite, and ancient. For countless generations, humans traced constellations, guided ships by starlight, and created stories from the glitter above. To our ancestors, the heavens were sacred and unreachable.

Today, that same sky hosts new travelers. Thousands of human-made companions move quietly in orbit, circling Earth like mechanical moons and carrying information instead of myths. They map continents, track storms, guide aircraft, connect distant voices, and watch our planet breathe.

Satellites reveal something remarkable. A place once ruled only by wonder has become a domain of human thought made physical. And like all profound advancements, it began with a single idea.

A Thought That Tried to Escape Gravity

Before rockets ever rose into space, a schoolteacher in Russia dared to think beyond the limits of his time. Konstantin Tsiolkovsky was not a pilot or an engineer. He was a philosopher shaped by mathematics.

While the world struggled to master flight, he asked a deeper question. Could machines travel beyond Earth and stay there, orbiting endlessly?

Through equations and imagination, he discovered that a craft launched with enough speed could break free from gravity. He spoke of artificial moons, long before such language became practical. His work did not shake the ground with noise, yet every satellite flying today still follows his principles.

From a quiet classroom came the foundation of the Space Age.

Sputnik: The Sound That Announced a New Era

In October 1957, a sphere the size of a beach ball (to be precise, 58 centimeters in diameter) left Earth behind. Sputnik 1 became the first artificial satellite and the first human-made object to circle the planet.

Its radio sent a simple repeating signal through the night.

Beep. Beep. Beep.

A calm heartbeat from space.

Millions heard it.

Nations felt it.

History shifted.

For the first time, humanity had placed a creation of its own among the stars.

The sky was no longer a distant ceiling. It had become a pathway.

From One Sphere to Thousands of Missions

Many decades later, orbiting machines serve purposes once unimaginable.

Communication satellites connect families across oceans.

Weather satellites warn us of storms before they form.

Navigation systems like GPS, GLONASS, and Galileo guide ships, aircraft, and even pedestrians.

Observation satellites monitor melting ice caps, study forests, and watch over farmlands.

They are human thoughts stabilized in orbit.

Dreams are programmed, launched, and given purpose.

A Silent Revolution Above Our Lives

The modern world depends on satellites almost without noticing.

Banks confirm transactions in milliseconds.

Pilots follow invisible paths through the sky.

Emergency teams track hurricanes hours before landfall.

Your smartphone's map is the work of space-borne helpers.

Innovation becomes true progress when people no longer marvel at it. It becomes a quiet part of daily life.

What Comes Next: A Crowded Sky

New inventors now bring a transformation of scale.

Thousands of satellites are launched each year.

Companies such as SpaceX build networks that bring the internet to remote mountains and deserts.

Students launch CubeSats, small enough to hold in two hands, yet powerful enough to conduct research once reserved for national agencies.

Space has opened to more minds.

The frontier is no longer distant. It is shared.

The Invisible Thread Between Earth and Sky

Satellites are not simply machines. They are evidence of a human desire to explore, to connect, and to understand. They extend our knowledge outward, one orbit at a time.

From Tsiolkovsky's chalkboard to Sputnik's heartbeat, to global networks that carry our voices and discoveries, satellites reflect the evolution of human imagination.

A Thought That Now Lives Among the Stars

The sky at night still appears unchanged.

No moving objects, only stars.

Yet above those stars lie thousands of our silent companions, circling continuously and connecting the world through unseen signals.

They are ideas in motion.

Curiosity given altitude.

A dream that reached upward and refused to fall back.

We once reached for the stars through stories.

Now we reach through invention.

The idea that began in a schoolteacher's mind has become a constellation of human progress.

A thought rose from Earth.

And it stayed in the sky.

Chapter 19
The Need for Traffic Control in Space

When Sputnik launched, humanity had just begun to understand space. Within a single human lifetime, we have transformed an empty sky into a complex system filled with machines, moving rapidly in curved orbits above our heads.

More than ten thousand satellites are active today. Over fifty thousand pieces of trackable debris race through the same space. The orbits around Earth have become busy lanes of motion, carefully planned yet inherently risky.

Satellites guide navigation, communication, weather prediction, and even farming. But as traffic increases, so does the need for coordination.

Unlike aircraft, there is no global controller managing orbital paths. Different organizations track objects independently. The United States Space Surveillance Network and the European Space Agency monitor potential collisions and send alerts. When danger appears, satellite operators adjust flight paths, sometimes with only hours to respond.

In 2019, the European Space Agency moved a satellite to avoid a collision with a Starlink satellite. Both machines were serving Earth, yet they almost collided in the sky meant to support them.

A new challenge rises with every launch. Space debris grows.

Dead satellites, unused boosters, and metal fragments travel at extreme speeds. Each piece is small, but dangerous. A single crash could create more fragments, leading to what scientists fear as the Kessler Syndrome, a chain reaction of collisions.

New ideas aim to prevent that future.

Robotic arms, nets, harpoons, and magnetic docking systems are being tested.

Laser technology could slow objects until gravity pulls them back to Earth.

ESA plans the ClearSpace-1 mission to remove debris.

Companies like Astroscale show how satellites might be collected and safely deorbited.

Engineers now design satellites that eventually burn up safely upon reentry.

Yet, cleanup progresses slowly because the sky is large and the debris is countless.

In only a few decades, we have filled the darkness above with extraordinary tools and the leftovers of our progress. The next step is responsibility.

We must protect the space we depend on.

We must learn not only to explore the universe, but to manage and care for it.

Chapter 20
The Thought of Robots – Machines That Think

We have always imagined life emerging from lifeless things. Ancient stories spoke of mechanical servants and artificial beings. Today, reality is catching up to myth. The idea of machines that think like humans began as philosophy and storytelling. Over centuries, it transformed into engineering, mathematics, and code. A spark of curiosity became a scientific pursuit.

The First Mechanical Minds

Long before the term robot existed, civilizations experimented with automata. Mechanical birds that fluttered, statues that rotated, water clocks that moved as if alive. The hero of Alexandria crafted devices powered by steam and counterweights. They were not intelligent, but they mirrored the idea that motion could be designed.

The word robot arrived centuries later. In 1920, Czech playwright Karel Čapek introduced it in his drama RUR. His robots were not metal. They were artificial workers created by science. Yet he planted a powerful question:

Would artificial beings always obey humans, or one day question them?

The Digital Foundation

Modern thinking machines emerged with the digital age. Alan Turing asked a defining question. Can a machine think like us? He proposed a test. If a computer can hold a conversation that convinces a human, does it deserve to be called intelligent?

This shifted the goal. It was no longer about gears and levers. It became about learning, decisions, and adaptation.

In the 1950s, the first industrial robot arrived. Unimate, created by George Devol and Joseph Engelberger, worked with precision on assembly lines. It was welded with steady accuracy. A machine had entered the workforce not only to assist, but sometimes to replace.

At the same time, the Dartmouth Conference of 1956 established artificial intelligence as a scientific field. John McCarthy and Marvin Minsky believed machines could eventually solve problems and plan like humans. That belief changed the direction of technology.

When Robots Started to Think

In the 1960s, Shakey the Robot appeared. A simple box on wheels, yet it could look around, process what it saw, and choose its next move. It did not just act on commands. It thought about action.

From that moment, robots and AI evolved together.

Sometimes quickly. Sometimes slowly.

Always forward.

Decades passed, and progress accelerated. Robots now assemble cars, navigate hospitals, and search disaster zones. Space missions rely on robotic explorers like those on Mars. Boston Dynamics created machines that run, jump, and balance. AI powers search engines, medical scans, translation tools, and voice assistants.

A machine does not need a human body to think. Software alone can solve problems faster than any brain.

The Questions That Follow Progress

With each advancement, unease grows.

What happens when a machine makes an independent decision that causes harm?

Who is responsible when jobs disappear due to automation?

If a robot writes poetry or expresses emotion, what remains unique about humans?

These questions are no longer theoretical. They shape how we design technology and laws. They push us to consider what consciousness means and whether intelligence must come from biology.

A Future That Arrives Quietly

Today, humanoid robots practice facial expressions. Self-driving systems reshape transportation. Homes anticipate our needs. The line between human and machine becomes thinner with every update.

The thought of robots began as an imagination.

Now it is part of our daily life.

And perhaps soon, machines will have thoughts of their own.

The journey continues. The creator and creation are learning from each other.

Chapter 21
How a Huge Airplane is Built

From Blueprints to the Sky

Every large airplane begins as a vision. A flying giant that can lift hundreds of people, carry heavy cargo, and cross entire continents in hours. Building such a machine is not just engineering; it is global teamwork and precision at the microscopic scale.

Designing the Sky Traveler

It starts in design labs. Aeronautical engineers simulate airflow over wings and fuselage using advanced software. They study lift, drag, and stress. Models are tested in wind tunnels. Every curve is shaped by physics and by the need for safety.

Nothing is left to chance. Thousands of virtual tests occur before the first real part exists.

Building Across the World

Once the design is approved, manufacturing spreads across nations.

Wings may be built in the United Kingdom.

The fuselage in Germany.

Tail sections in Spain.

The cockpit in France.

These enormous pieces travel by specialized ships and aircraft to assembly plants.

Putting the Aircraft Together

On the production line, humans and machines work side by side.

Massive cranes lift huge structures into position.

Robots drill holes and install fasteners with remarkable accuracy.

Technicians weave kilometres of wiring, connect hydraulic systems, and install digital controls.

Every bolt is checked. Every joint inspected. Safety guides every move.

Then the engines arrive, tall as houses and powerful enough to lift hundreds of tonnes into the air. They compress air, ignite fuel, and create thrust that allows flight through a thin atmosphere.

With engines attached, the aircraft begins to look alive.

The Tests Before Flight

Before passengers ever sit in a seat, the aircraft undergoes intense testing. Wings are bent nearly to their maximum flexibility.

Emergency systems are triggered.

Pilots take test flights and evaluate behaviour from takeoff to landing.

Only when every standard is met is the airplane certified to fly the public.

The Achievement in Motion

A modern plane is a flying city.

Inside it are the ideas, skills, and labor of thousands of engineers, designers, and workers from different places around the world.

Every successful flight is a reminder.

Humans learned to master the sky not through luck, but through cooperation, science, and determination.

Chapter 22
The Bird That Banged Its Head and Taught Us How to Survive Crashes

A bird slams its head against solid tree trunks 20 times per second, each hit at a force over 1,200 times the force of gravity. It does this thousands of times a day. And yet, it never gets a concussion.

We call it a woodpecker.

Nature calls it an unbreakable engineer.

Long before humanity built cars, airplanes, or contact sports, this bird had already evolved the perfect shock-absorbing system. A system that protected its brain, preserved its life, and kept it tapping rhythmically against bark, season after season.

But how?

When scientists slowed down the woodpecker's head-banging with high-speed cameras and CT scans, they discovered a masterpiece of bioengineering. Four major components formed its natural crash-helmet:

Beak Design – The upper and lower beaks are slightly unequal in length, reducing direct force transmission to the brain.

Spongy Bone – Behind the beak sits a dense but spongy bone matrix, especially at the front of the skull, that gradually dissipates shock energy.

Hyoid Bone Wrap – A long bone called the hyoid wraps around the skull like a seatbelt, offering structural reinforcement and energy redirection.

Minimal Cerebrospinal Fluid – Less fluid means the brain has less room to jostle, reducing internal impact during rapid motion.

For years, this remained a curiosity of biology. Then came a turning point when researchers at Beihang University in China, inspired by the woodpecker's resilience, began rethinking how humans handle shock, especially in aviation and safety.

The result? A revolution in impact protection.

Black box flight recorders, those near-indestructible memory devices inside airplanes, were redesigned with woodpecker-inspired cushioning layers. Just as the bird's skull disperses energy in stages, engineers began crafting multilayered casings, each absorbing a fraction of the impact, making even catastrophic crashes survivable for data.

Helmets, too, began to change. Sports helmets, military gear, and even future astronaut headgear drew on the idea of layered, uneven structures with flexible shock zones mirroring the beak-to-brain buffer the bird evolved. Football helmets are now being developed with inner lattices modeled after the hyoid wrap, reducing rotational forces that cause concussions.

All this from a bird tapping away quietly in the woods, unaware that it was solving a modern problem that would save human lives decades, even centuries, later.

Nature never shouts. But it leaves hints. In feathers, in bones, in silent rhythms. The woodpecker didn't mean to become our crash-test model. But in doing what it had always done, it showed us something profound: the best shock absorbers are not always synthetic; they are often alive, adaptive, and honed by evolution.

Sometimes, the loudest invention begins with a simple knock.

Chapter 23
Bullet Trains – Inspired by the Kingfisher's Beak

There are moments when a quiet observation leads to a loud revolution.

The history of Japan's Shinkansen, the Bullet Train, began with such a moment.

In the late twentieth century, engineers had created trains of remarkable speed.

But each time these trains exited a tunnel, a violent sonic boom shook homes and disturbed communities.

The cause was a sudden air pressure burst against the train's blunt nose.

Mechanical solutions failed.

Progress needed a new perspective.

Eiji Nakatsu, an engineer and devoted birdwatcher, found that perspective in nature.

While observing a kingfisher dive, he noticed something extraordinary.

The bird pierced from air into the water smoothly with almost no splash.

Its slender beak sliced through two different environments as if they were one continuous space.

Nakatsu saw a solution that others overlooked.

If a bird could move between air and water without disturbance, perhaps a train could move from a tunnel to open space just as quietly.

He redesigned the front of the Shinkansen to mimic the kingfisher's beak.

A graceful aerodynamic curve replaced the old blunt form.

The result amazed the world.

The train became faster.

The noise faded into a soft rush.

Energy use dropped significantly.

Nature had whispered the answer.

Engineering finally listened.

But recognition arrived slowly.

During the train's celebrated launch, Nakatsu was asked not to attend.

His nature-inspired approach was considered too unconventional.

Industrial leaders chose prestige over the quiet thinker who made the breakthrough possible.

While cameras flashed and crowds applauded, he watched from a distance.

Invention had transformed the world, yet its creator remained outside the spotlight.

This silence is familiar in history.

Those who listen deeply are often noticed last.

Time corrected this silence.

Years later, Nakatsu was honored as Chief of Japan's Space Transportation System development.

His path had always pointed upward, from the water diving kingfisher to the stars.

Nature has spent millions of years refining efficiency.

Humanity advances when it chooses to observe rather than conquer.

The owl taught us silent flight.

The shark taught us smooth movement through water.

The kingfisher helped us glide through tunnels at breathtaking speed.

Every one of these is a partnership with nature, not a victory over it.

The Bullet Train is more than a technological triumph.

It is a symbol of humility.

A reminder that human progress often begins with quiet attention.

When we listen to the world around us, the future moves forward faster and with far less noise.

Chapter 24
Drone Technology – Eyes in the Sky

There was a time when only birds, astronauts, and daring pilots could look down upon the world from above. Today, small machines with spinning wings share that view. Drones are silent observers in the sky. Farmers use them to study fields. Filmmakers use them to capture beauty. Firefighters use them to find heat hidden in smoke. Generals use them to watch, and sometimes to strike.

Their roots, however, began in war. During World War I, engineers experimented with unmanned aircraft loaded with explosives. These early "aerial torpedoes" rarely succeeded, but they introduced a powerful idea. Machines could enter danger, so people would not have to.

By the Vietnam era, drones were no longer experiments. The United States deployed them for reconnaissance, sending them over hostile territory to gather intelligence without risking pilots. In later conflicts, including Afghanistan and Iraq, drones such as the MQ-1 Predator and MQ-9 Reaper changed the battlefield. They not only watched, but they attacked. Guided from remote locations, operators could conduct missions from thousands of kilometers away. This raised urgent ethical questions. If the trigger is pulled from another continent, where does accountability live?

War pushed technology forward, as it often does. GPS became accurate. Cameras became sharper. Sensors became smaller. Drones soon stepped out of military camouflage and into civilian hands. In the early 2000s, they were reborn as tools of peace. They monitored crops, delivered supplies, searched disaster zones, and expanded creativity.

Modern drones are engineering achievements. Gyroscopes keep them level. GPS guides their paths. Artificial intelligence helps them avoid obstacles. They survey melting glaciers, deliver medical aid to isolated communities, track wildlife, and create light shows that rewrite the idea of fireworks.

Yet with this growth comes new concerns. Crowded skies increase the risk of collisions. Privacy becomes easier to invade. Criminals find ways to misuse technology. Nations now race to define rules for this invisible highway above our heads.

What began as a wartime experiment has lifted human perspective. From farmers to filmmakers, from rescuers on Earth to explorers on Mars, drones extend our vision in every direction. The sky, once distant and unreachable, is now a place humans can touch through machines.

Chapter 25
Elon Musk – The Mind Rewiring Tomorrow

Elon Musk is more than an entrepreneur. He is a force pushing the boundaries of what our future could become. Born in Pretoria, South Africa, in 1971, he asked questions that felt too large for childhood. Could humans live among planets? Could energy be clean and infinite? Could technology erase the limits that slow us down?

His journey began in software. At twelve years old, he coded and sold a video game. A quiet signal of the ambition inside him. Yet software was only his first horizon.

PayPal changed how money moved in the digital age. But while others celebrated success, Musk looked upward. His attention shifted to the stars.

SpaceX, founded in 2002, challenged the rule that only governments belonged in space. Musk envisioned rockets that could return to Earth and fly again. Early launches failed. Rockets fell into the ocean. Funding nearly ran out. Most would have stopped. Musk continued.

The breakthrough came with Falcon 9, a rocket that delivered payloads into orbit and then landed upright on Earth. Space travel became less expensive and more frequent. SpaceX opened new doors for satellites, scientific missions, and dreams of reaching Mars.

On Earth, Tesla transformed the automobile from a gasoline machine into a clean, powerful, and intelligent device. Electric cars became desirable. Tesla's advanced battery storage, self-driving technology, and over-the-air updates made cars better long after delivery.

SolarCity pushed renewable energy into homes and neighborhoods. The Boring Company tested new ways to escape traffic by building tunnels beneath cities. Neuralink explored possibilities of linking the brain and computer, aiming to restore lost abilities and extend human cognition.

Each company seems different, yet all share one thread. Musk builds systems. Ecosystems of technology that connect energy, transportation, communication, and the future of life itself.

He has critics. Some see him as reckless. Others as a visionary. His work ethic is relentless. His expectations are high. His risks are enormous. But innovation rarely grows in safe soil.

Elon Musk is an example of how ideas can reshape reality. He invites humanity to think beyond comfort and habit. He reminds us that progress is often difficult. It is imperfect and uncertain. Yet it is the force that turns imagination into the world we wake up in.

Epiphanies from This Wave

PART 2 – Machines of the Mind

— *A child's bamboo toy helicopter spun skyward centuries later, and that playful thought became the roar of airplanes piercing the clouds.*

— *The hiss of steam from a kettle seemed ordinary until it became the engine that pulled nations into the age of industry.*

— *Franklin's kite, trembling in the storm, caught a bolt of lightning. That spark did not power cities, but it taught humanity that darkness could be challenged.*

— *The kingfisher dived without a splash and its beak taught engineers to carve trains that sliced the wind in silence.*

— *The carriage shed its horses, wheels turned with fire, and suddenly distance collapsed, freedom rolled on four wheels.*

PART 3

Digital Tsunami

From Computers to Consciousness

Chapter 1
Computers: How Thought Became Machine

From Turing's Mind to a World That Thinks for Us

I step into the cool evening air. Streetlights hum above me. The quiet glow of screens flickers behind windows. Cars move by being guided by sensors and circuits. Everywhere I look, machines are thinking.

Not long ago, thought lived only in human minds, scribbled in notebooks or whispered between scholars. Then came a shift. A moment in history when thought leapt out of the brain and into metal. Equations became power. Logic turned into a language that machines could speak.

Alan Turing
Photograph by
Elliott & Fry, 1951

We built a world where ideas control engines and algorithms shape entire cities. How did we go from pure theory to a planet that automates itself?

The First Thought of a Machine: Turing's Invisible Idea

Before computers existed, they were imagined.

And the imagination belonged to one man, Alan Turing.

In 1936, Turing described something extraordinary. A machine that lived only in concept. It could read symbols, follow rules, and make decisions. A process, not a device. He showed that logic could be mechanized. That reasoning could be engineered.

This theoretical model became known as the Turing Machine. It proved that any calculation the human mind could perform step by step, a machine could do too.

No circuits yet. No screens. Just the seed of possibility.

The world now needed to build what Turing had already seen.

When Thought Took Shape: The First Computers

The earliest computers were not elegant. They filled entire rooms. They rattled and hummed like industrial beasts. Wires tangled like nervous systems. Lights blinked like uncertain eyes.

In 1941, Konrad Zuse created the Z3. It was programmable and digital, built from the determination of one man working largely alone.

In 1945, ENIAC erupted into life in the United States.

Fully electronic.

Thousands of calculations per second.

A speed no human could touch.

These machines did not look intelligent. But they did something revolutionary. They turned thought into a physical force.

The Revolution Inside: From Vacuum Tubes to Transistors

The future of computing did not arrive with noise. It arrived quietly in 1947, when three scientists at Bell Labs invented the transistor.

Small. Fast. Cool to the touch.

It replaced bulky tubes and made the computer shrink from a giant monster into a dependable companion.

By the 1960s, computers left scientific labs and entered business offices. The machine was moving closer to everyday life.

A Computer for Every Desk: Microprocessors and the Home Era

In 1971, Intel introduced the 4004 microprocessor.

For the first time, the brain of the machine could sit on a tiny piece of silicon.

Suddenly, computing could be personal.

In 1977, the Apple II arrived.

Steve Wozniak engineered its guts.

Steve Jobs envisioned its purpose.

Families brought computing into their living rooms. Students learned programming as naturally as writing. With a floppy disk in your hand, you could hold knowledge and creativity in a pocket-sized container.

The computer became a household member.

From Tool to Partner: The Age of Automation

The 1980s introduced friendly interfaces.

IBM PCs. The first Macintosh.

The mouse replaced switches. Icons replaced command lines.

Then came the true transformation.

Connection.

In the 1990s, computers began talking to each other.

The internet emerged as a digital nervous system.

Machines could learn what we liked, guide planes, diagnose diseases, manage farms, and help doctors save lives.

We no longer command computers all the time.

They began suggesting what we should do next.

Automation quietly entered our daily decisions.

Where We Stand: A World of Thinking Machines

Today, computing is everywhere.

In pockets. In vehicles. In factories. Inside medical devices.

It schedules, translates, predicts, and steers.

Cloud networks blur physical boundaries.

Artificial intelligence learns patterns we cannot see.

Machine learning builds models even their creators struggle to explain.

On the horizon is quantum computing, using probabilities instead of bits. A place where logic meets uncertainty and speed becomes almost unimaginable.

The machine keeps evolving faster than we can.

Full Circle: When Thought Became Real

It began with a young mathematician imagining a process.

A mind, a tape, and a set of rules.

Now his idea lives in every chip and every code that runs our lives.

We built machines in our image, logical and tireless.

We handed them tasks once possible only for humans.

Behind every server, every search bar, every digital assistant,
a trace of Turing remains.

His machine hums quietly
in the background of the modern world.

Chapter 2
The Man Who Saw Through the Machine

War engulfed the world.
Armies clashed. Walls fell. Millions suffered.
Yet in a quiet room at Bletchley Park, a single man sat with a pencil and thought.
While others fought with rifles, he fought with logic.
His name was Alan Turing.
From childhood, he asked questions others ignored.
Why does thought work?
What are the limits of the mind?
His teachers found him strange.
Mathematics found him essential.

A Blueprint for the Future

In 1936, Turing published a paper that redefined possibility.
He imagined a theoretical machine that could follow instructions on a tape, step by step.
This model would become the foundation of computer science.
He was in his twenties.
The world had not yet caught up to him.

A War to Decode: The Enigma Challenge

During World War Two, Germany communicated through the Enigma machine, creating millions of unpredictable codes every day. Allied leaders feared that breaking it was impossible.

Turing did not accept the impossible.

He helped build the Bombe, a machine designed to break the Enigma's secrets faster than the battlefield could change.

Each day, it read the enemy's hidden intentions.

It saved ships from destruction.

It shortened the war.

Historians estimate that millions of lives were preserved because Turing's logic defeated German secrecy.

Yet his success remained hidden from public praise.

The war moved forward.

Turing returned to his work with quiet resilience.

Can a Machine Think?

After the war, he focused on a deeper challenge.

Could a machine think like a person?

In 1950, he introduced the Turing Test, asking a question that continues to shape artificial intelligence:

If a machine could hold a conversation so well that a human could not tell the difference, is it thinking?

It was not science fiction.

It was a philosophical challenge that still has no final answer.

The Price of Genius

Despite saving nations, Turing was criminalized.

He was a gay man in a society that punished love.

He faced humiliation and forced medical treatment.

In 1954, he died at the age of forty-one.

A half-eaten apple lay beside his bed.

Some interpreted it as a tragic choice.

Others believe it was an accident.

We may never know.

But history remembers what the world once tried to erase.

Justice Arrives Late

Decades passed.

Computers multiplied.

Encryption strengthened.

AI emerged.

Finally, the world recognized the man who made it possible.

In 2009, the British government issued an apology.

In 2013, the Queen granted him a pardon.

But the truest honour lives elsewhere.

Every algorithm.

Every code-breaking tool.

Every conversational machine.

They all speak a language Turing helped create.

He did not only break codes.

He broke the limits of thought.

Chapter 3
The CPU – A Modern Mimic of the Human Brain

Inside every computer, tiny pulses of electricity make choices in a language we designed. A thought emerges: how closely does this machine resemble our own mind?

From the beginning, the connection was intentional. Early computing pioneers such as John von Neumann studied the human brain and used it as a model. The brain receives inputs through the senses, processes them with neurons, and then produces actions. Engineers mirrored this idea.

The result became the CPU, the Central Processing Unit.

It receives instructions, decodes them, processes them, stores the results, and then sends out the output.

Circuits fire like neurons.

Memory modules work as the brain's hippocampus, storing both long-term data and momentary thoughts. The control unit focuses tasks, similar to our prefrontal cortex, guiding attention and decisions.

Von Neumann explained these parallels in his book The Computer and the Brain. He noted important differences, too. The human brain handles thousands of signals at the same time, while early computers handled tasks step by step. Yet the idea remained powerful. The more we understood the brain, the more we refined our machines.

Today, the line between biology and technology becomes thinner. Neuromorphic computers are built to behave like neurons, not just look like them. They respond, adjust, and learn. Artificial intelligence systems recognize faces, understand language, and generate new ideas. They run on networks inspired by human thinking.

In these systems, we see a reflection. It is not our biology we copied. It is our method of reasoning, our problem-solving structures, our curiosity transformed into circuits.

We did not simply invent the CPU.

We revealed how deeply we wanted to understand ourselves.

Chapter 4
The Countess Who Imagined the Future

In 1833, a young woman entered a room full of gears and polished brass. Most saw a machine of numbers. She saw a mind waiting to awaken. Her name was Ada Lovelace.

Ada Lovelace
Portrait by
Alfred Edward Chalon, 1838

Lovelace

Born Augusta Ada Byron, she carried the legacy of her father, the poet Lord Byron. He left before she could remember him.

Fearing Ada would inherit his emotional storms, her mother trained her in mathematics, logic, and careful thinking.

Yet Ada held onto imagination. She called her approach "poetical science". To her, numbers were not cold. They could breathe, express patterns, and create beauty.

At seventeen, she met Charles Babbage, a mathematician designing a mechanical computer he called the Analytical Engine. Babbage built machinery. Ada saw possibilities.

In 1843, she translated an academic paper about the Engine. She added her own notes, expanding a simple translation into a visionary document three times longer than the original. Within those notes was something new: a method for the Engine to process a sequence of operations. It is recognized today as the first computer algorithm ever written.

At seventeen, Ada Lovelace met Charles Babbage, a professor and inventor who was struggling to bring his mechanical ideas to life. He called his design the Analytical Engine. It was a machine built with brass, wheels, and punched cards, yet its purpose reached far beyond machinery. Babbage imagined a device that could follow instructions and transform input into meaningful results.

Charles Babbage

Charles Babbage
Daguerreotype by
Antoine Claudet, c. 1847–1851

Many viewed the Engine as an advanced calculator.

Ada saw something more.

Where others saw metal parts, she recognized the outline of a future mind.

In 1843, she translated an Italian paper discussing Babbage's invention. Most translators would have stopped there. Ada continued.

Her notes expanded into a profound commentary nearly three times longer than the original text. Inside those pages, she described the first sequence of coded steps ever intended for a machine to follow. That sequence would later be recognized as the first computer program in history.

Ada did not stop at algorithms.

She predicted how deep the machine's potential could grow.

The Analytical Engine, she wrote, might one day compose music, draw art, and manipulate symbols. It could become a partner to human creativity. Those ideas were so advanced that even scientists of her day struggled to accept them. The world would wait more than a century before artificial intelligence would begin to prove its worth.

Ada understood something extraordinary.

A calculating engine does not only handle numbers.

Numbers can represent anything.

Letters, images, sounds, even human thought.

Once a machine could work with symbols, it could participate in thinking itself.

Her life was brief. She died at thirty-six, the same age at which her father, Lord Byron, passed away. Yet the impact of her ideas continues to expand. She changed the future not by building a machine, but by revealing what machines could become.

Today, a structured programming language used by NASA and the United States Department of Defense is named in her honor. The name is simple: ADA.

Every line of code written since then carries a small echo of her vision, a reminder that mathematics and imagination are not enemies. They are partners in discovery. Somewhere in the quiet logic of software, her poetical science still speaks, guiding machines toward creativity.

Chapter 5
The Invention of the Internet

A letter once took weeks to cross oceans. Messages depended on ships, storms, distance, and luck. The telegram brought speed yet stripped away emotion, turning feelings into short commands. Telephones brought voices closer, but only when both ends stayed connected. Then a question emerged. *What if* information could move faster than any ship or cable? *What if* communication could think faster, respond faster? *What if* it could travel at the speed of thought?

That question gave birth to the internet. A bridge built not from bricks or steel, but from invisible pathways that carry ideas, knowledge, emotions, and entire worlds.

ARPANET: When Machines First Spoke

In 1969, four computers located at UCLA, Stanford, UC Santa Barbara, and the University of Utah talked to each other for the first time. The network was called ARPANET, funded by the United States Department of Defense. The Cold War had taught an important lesson. Information must survive even if cities fall.

Packet switching became the strategy. Messages were broken into small units, sent across different routes, and reassembled at the destination, allowing communication to continue even if part of the network failed.

During the first test, a computer at UCLA attempted to log into a machine at Stanford. The system crashed after the first two letters. The screen displayed only "LO". Short for "LOGIN". Also, a quiet declaration that said, "Lo, behold." A new age had begun.

One Symbol That Changed Everything

In 1971, Ray Tomlinson sent the first email. It was simple. It may have contained nothing more meaningful than a row of random keyboard characters. Yet it included a symbol that would change communication history. The @ symbol linked a username to a machine. A small mark found its purpose, connecting people across distance.

Email soon introduced intimacy into machines. By the 1990s, someone in Lagos could write to someone in London as naturally as sending a note across a classroom. Communication grew faster, personal and global, all at once.

Making the Computer Personal

Networks needed people. People needed machines they could actually use.

In 1975, Bill Gates and Paul Allen focused on writing software for a personal computer. Meanwhile, Steve Jobs and Steve Wozniak built new machines inside a small California garage. The result was the rise of personal computers like the Apple II.

Computers moved out of laboratories and entered homes, offices, and schools. The internet was no longer a restricted tool for researchers. It was becoming a part of everyday life.

Tim Berners Lee: What If It Were All Connected

In 1989, at CERN in Switzerland, computer scientist Tim Berners Lee had a revolutionary thought. *What if* all the information scattered across networks could be connected seamlessly? *What if* humans could jump from one idea to another as easily as neurons fire inside a brain?

He created:

- HTML to structure information
- HTTP to transport it
- URLs to locate it

By 1991, the World Wide Web emerged. Anyone with a computer could explore a new digital landscape.

Web browsers like Mosaic and Netscape followed. Information turned visual. Pages became colorful and interactive. A child could explore knowledge instantly. A grandmother could send photos across continents.

The internet became not only useful, but alive.

Organizing the Digital World

As the web expanded, a new challenge formed. How do we find anything inside this growing ocean of data?

Two Stanford students, Larry Page and Sergey Brin, believed not all information was equal. Some sources deserved more trust and visibility. They built Google. It was more than a search engine. It ranked ideas.

Other companies transformed the digital space:

- Amazon created a limitless bookstore
- eBay connected global buyers and sellers
- Facebook allowed individuals to share life itself
- YouTube turned audiences into creators

The internet evolved into a marketplace, a library, and a stage all at once.

The Price of Connection

Connectivity brought contradictions.

The internet democratized knowledge while making truth harder to protect.

It created friendships while enabling harassment.

It gave people a voice while sometimes amplifying chaos.

Technology accelerated. Satellites brought networks worldwide. 5G boosted speed. Artificial intelligence learned to predict what we think, sometimes before we think it.

Today, devices communicate constantly. From wristwatches to refrigerators, machines are no longer silent. They listen, record, calculate, and respond.

The Thought That Connected Us All

The internet began not with a wire, but with a worry. *What if* communication fails? Then a curiosity. *What if* machines help each other? Then a dream. *What if* every mind could reach every other mind instantly?

With two letters glowing on a screen, "LO", humanity stepped into a networked future. A future imagined by thought, built by thought, and still evolving through thought.

Chapter 6
Human Language and Computer Language
A Surprising Resemblance

We speak, and others understand.

We type into computers, and they respond.

Two very different systems, yet both built on a remarkable foundation: language.

At first, comparing English or Tamil to Python or Java sounds unreasonable. One evolved naturally over thousands of years. The other was intentionally engineered in modern laboratories. One is emotional and expressive. The other is precise and literal.

Yet both have structure. Both follow the rules. Both attempts to carry meaning from one mind to another.

Structure and Precision: The Shared Architecture

Human languages rely on grammar to ensure clarity. A misplaced comma can shift meaning or tone. In computer programming, a single missing semicolon can cause a program to fail entirely. In both systems, discipline is essential.

A sentence that is poorly structured confuses the listener.

A line of faulty code confuses the machine.

George W. Smith, in his work Computers and Human Language, explains how computers borrow tools from human linguistics to process information. Parsing, semantics, and contextual interpretation all originate from how humans use language. Computers analyze sentences by splitting them into smaller parts, just like a student breaking down clauses in school.

Machines do not understand words the way we do, yet they rely on the same principles to function.

Evolution: Growth Through Change

Languages are not fixed.

They adapt and transform.

Human languages adopt slang, dialects, and shifting expressions as cultures evolve. Programming languages also change. Some become obsolete. Others merge, improve, or grow to be faster and more flexible.

What once belonged to universities and research labs now shapes global industries. Programming has become a universal literacy.

And in a remarkable reversal, we are now teaching computers to interpret our natural speech.

Voice assistants.

Real-time translation.

Chatbots that attempt conversation.

These systems rely on deep learning models trained on billions of words. They no longer follow only rigid instructions. They learn patterns, structure, and meaning through exposure and experience.

Analogy: The Core of Human Thought

Cognitive scientist Keith Holyoak, in The Human Edge, highlights analogy as a hallmark of human intelligence. We understand new concepts by connecting them to familiar ones.

Modern artificial intelligence does something similar.

It identifies patterns.

It draws connections.

It learns relationships within vast collections of text.

Whether a parent soothing a child with gentle words or a programmer constructing a loop to automate a task, the underlying force remains the same. Language carries intention and intelligence.

One Impulse, Two Forms of Expression

Human language is emotional and intuitive.

Computer language is logical and unwavering.

Yet both emerge from the same desire inside us:

To communicate.

To instruct.

To be understood.

The resemblance between these two forms of language is not a coincidence. It is a mirror. Computers speak because we taught them how. Their languages exist because we first created our own.

And as technology continues to grow, the boundary between human expression and machine communication becomes less defined. We are not only shaping how computers talk.

We are shaping how we think.

Chapter 7
Mobile Phones – A Whole World in Your Pocket

There was a time when the telephone belonged to a single place.

It was a solid device, fixed to a wall or resting on a wooden table, waiting patiently for someone to lift its receiver. If the person you wished to reach was not at home or not near that phone, the message simply vanished. Words remained confined to the mind, unheard and unresolved.

Then a new possibility emerged.

A bold question formed in human thought.

What if the telephone could travel with the person?

The year was 1973.

On a crowded sidewalk in New York City, an engineer named Martin Cooper of Motorola raised a bulky device to his ear. Roughly the size of a brick and weighing more than two pounds, it was far from elegant. He pressed a button and placed the first public mobile phone call. The call was not made to a friend, but to a rival at Bell Labs. The battery failed within thirty minutes, yet something irreversible had occurred. Communication was no longer anchored to furniture. It had gained mobility.

Phones gradually learned to follow us.

Throughout the 1980s and 1990s, they became smaller, lighter, and increasingly affordable. What began as an exclusive tool for a few evolved into a common companion for many.

For millions, that moment arrived with the Nokia 3310.

Sturdy. Reliable. Remarkably durable.

It became a cultural symbol of early mobile life, introducing vast numbers of people to portable communication and to a simple, addictive game called Snake. This device marked a transition. Phones were no longer limited to conversation. They were beginning to integrate into everyday routines.

Yet the human mind rarely rests on achievement. Progress invites further ambition.

If a phone could move, why should it only speak?

The Arrival of the Smartphone

A new idea took shape.

Make the phone intelligent.

BlackBerry moved first, placing email directly into the hands of professionals, government officials, and business leaders. Communication accelerated. Work became untethered from offices and desks, following people wherever they went.

Then came 2007, a defining moment in modern history.

A man in jeans and a black turtleneck stepped onto a stage and declared that Apple was about to reinvent the phone.

The iPhone arrived not merely as a communication device, but as a convergence of functions. It was a camera, a music player, a library, a notebook, a pocket computer, and a gateway to the internet. One object replaced many.

The transformation went beyond communication.

It reshaped daily life.

Today, the device in your pocket connects you to satellites orbiting Earth. It remembers important dates, navigates traffic, monitors heart rhythms, and predicts preferences. It informs, entertains, assists, and occasionally challenges how you see yourself and the world around you.

A student in a remote village can attend a global classroom.

A physician can monitor a patient's heart from miles away.

A family separated by an ocean can share the same smile at the same instant.

Entire industries rose from this innovation:

- App development
- Mobile commerce
- Mobile banking
- Social media
- Digital health technology
- Mobile gaming, now generating more revenue than Hollywood

A phone is no longer a device.

It is a platform for human activity.

The Paradox of Constant Connection

Yet progress always brings a shadow.

With the world in our pockets, we often forget the world around us.

Screens sometimes replace presence.

Notifications interrupt thought.

Private life becomes data.

Every tap and every scroll is quietly recorded and analyzed.

The more connected we became, the more attention became a commodity.

The Future in Motion

Still, the idea evolves.

Mobile phones now fold, while some no longer look like phones at all.

Smart watches listen to our pulse.

Voice assistants respond when called.

Artificial intelligence learns our habits and completes our sentences.

And a new question quietly grows.

What if the device disappears entirely?

Perhaps the future phone will be a pair of glasses that projects information into our vision.

Perhaps direct signals from the brain will replace physical tapping on a screen.

Perhaps the next phone will be so integrated that we forget it exists.

What if your next phone is you?

From Martin Cooper's sidewalk call to advanced wearable technology, the mobile phone has not simply followed human progress. It has guided it.

Connected us. Expanded us.

Rebuilt societies and expectations.

And it all began with a simple thought:

Stay connected wherever you are.

Chapter 8
Steve Jobs: The Visionary Who Saw Tomorrow Today

Steve Jobs never walked with the crowd. Even as a child, he moved to the rhythm of his own curiosity, questioning rules, creating new ideas, sometimes clashing with those around him, yet always dreaming far beyond what others could see.

Born in 1955 in San Francisco and adopted soon after, Jobs grew up in Silicon Valley. The region was transforming into a hub of technology and restless ambition. That environment fueled him. He did not fit the traditional model of success. School felt too confined for his mind, so he stepped in and out of college, collecting only what he found meaningful. A calligraphy class shaped Apple's elegant typography. Zen philosophy influenced his simplicity and calm design. His pursuit of perfection, however, often crossed into obsession.

Building Apple and Building a Vision

In 1976, inside a modest garage, Jobs and his friend Steve Wozniak built the Apple I. What seemed like a simple machine became the first step in a deeper vision. Jobs wanted computers to be personal, friendly, and beautiful. Apple grew quickly, powered by bold ideas like the Macintosh with its graphical interface and intuitive controls.

But Jobs' personality was a double-edged sword. His sharp temper and unrelenting expectations inspired extraordinary results but strained many relationships. In 1985, he was forced out of the very company he helped create. The world saw a fall. Jobs saw a chance to build again.

Exile, Reinvention and Return

He founded NeXT, a company focused on creating the perfect computer for professionals. He acquired Pixar and helped transform animation into a new form of storytelling that blended technology with emotion. Meanwhile, Apple lost direction and struggled to compete.

In 1997, Apple called Jobs back. The company was close to collapse. What followed changed history. The iMac simplified the computer. The iPod placed a thousand songs in every pocket. The iPhone redefined communication. The iPad turned screens into personal notebooks and studios.

Jobs did not just create products. He created desire. He understood that technology becomes powerful when it touches human experience. His keynote presentations, which people called the "reality distortion field," made audiences believe the future was already in their hands.

The Private Man Behind the Public Legend

Behind the staged confidence was a private human being. He confronted fear, fought illness, and guarded his personal life carefully. His battle with cancer eventually ended his life in 2011. Yet his impact continues wherever light glows from a screen or a device responds to a touch.

Steve Jobs taught us to look at the world not only as it is, but as it could be. He believed technology was not metal and circuits. It was imagination, design, and connection. He changed how we communicate, how we create, and how we think about possibility.

His legacy lives in every innovation that dares to merge art with intelligence.

Steve Jobs did not simply predict tomorrow. He built it.

Chapter 9
When a Rock Pulled a Nail

A simple rock, dark and unremarkable,

once pulled a metal nail toward itself.

That tiny movement changed everything.

Without that Lodestone, the compass would not have pointed north.

Without compasses, sails would wander without direction.

Without ships crossing oceans, there would be no worldwide trade.

No discoveries. No wireless communication.

No Faraday experimenting with coils.

No radio waves traveling through the air.

No phones in our pockets.

No satellites guiding us through the night.

All because one stone responded to an invisible force.

Magnetism led to electromagnetic theory.

Electromagnetic theory led to a global connection.

Humanity moved faster the moment a rock tugged a nail.

There was a time when the world felt enormous and slow.

Messages traveled by foot, horse, pigeon, or ship.

News arrived late. Help arrived too late.

Sometimes a delay changed destinies forever.

The desire to shrink distance lingered in the human mind.

A thought that grew gradually.

From a magnetic stone to an electric spark.

From silent codes to the sound of voices.

From wired signals to wireless waves.

The First Breakthrough: Making Silence Speak

Before a voice could travel, a message had to move.

Samuel Morse did not enter communication through science, but through sorrow.

In 1825, while he worked as a painter in Washington, he learned of his wife's sudden illness only after days of travel.

By the time he reached home, she had already passed away.

Distance had stolen time from him.

He could not accept that barrier.

Morse dedicated himself to solving the problem.

In 1844, he tapped the first public telegram:

"What hath God wrought?"

A short sentence with enormous impact.

The telegraph turned silence into a coded language of dots and dashes.

Cities gained new arteries.

Continents were stitched together by cables under the sea.

But the telegraph spoke in code, not in human emotion.

The heartbeat of a voice was still missing.

The Second Breakthrough: Wires Learn to Speak

Alexander Graham Bell entered history not as an engineer first, but as someone trying to help people hear.

His father taught the deaf.

Bell wanted to restore sound for those who lived without it.

He believed that the same wires that carried codes could carry a voice.

In 1876, during an experiment, his assistant Thomas Watson heard the first transmitted words:

"Mr. Watson, come here, I want to see you."

For the first time in human history, a voice traveled through a wire.

No translator. No training.

Just the natural sound of one person calling another.

The telephone spread quickly.

Businesses, homes, and entire nations found new ways to communicate.

The world became smaller.

Quieter distances. Louder connections.

From Wires to Air

While telephones transformed life, another bold idea emerged.

What if messages could move through the air, without any wires at all?

In the late 19th century, Guglielmo Marconi demonstrated that electromagnetic waves could leap through open space.

The radio was born.

Communication no longer needed a physical path.

Sound could ride on invisible currents.

A century later, a greater transformation arrived.

Not only sound, but information itself became wireless.

The digital age accelerated.

The Internet escaped desktops and demanded freedom of movement.

People wanted to stay connected everywhere.

Not tied to walls or ports.

The Rise of Wi-Fi

Wi-Fi, short for Wireless Fidelity, answered that demand.

Its early years were imperfect.

Bulky routers. Weak signals. Slow speeds.

Yet the idea was unstoppable.

With improved standards such as IEEE 802.11, engineers unlocked new freedom.

Homes, cafés, parks, airports, and trains all became places of connection.

A student under a tree could reach the world with a single click.

The physical chains of communication began to disappear.

Mobility Becomes a Way of Life

Meanwhile, phones evolved.

Voice became text.

Text became photos, videos, and live calls.

Bluetooth allows devices to communicate with one another.

Satellites extended their reach to remote mountains.

The Internet transformed speech into data and freed it from landlines entirely.

Connection became constant.

Not an action, but a state.

A whisper in New York can be heard in Nairobi.

A laugh in Tokyo can appear instantly in Toronto.

Distance lost its power.

From One Stone to the Stars

It all began with a silent pull.

A rock attracting iron.

That pull became a question.

The question became a theory.

The theory unlocked the electromagnetic world.
We learned to speak in sparks.
To send sound across oceans.
To turn ideas into invisible waves.
From Morse tapping a key
To bell, calling a friend
To radios echoing across nations
To wireless signals orbiting the planet
We are now a species connected not by land, but by signal.
And none of this human transformation would have begun
If a rock had not once pulled a nail.

Chapter 10
Guglielmo Marconi – The Man Who Whispered Through the Air

It did not begin with lightning, but with a question.

What if messages could travel without wires?

What if the invisible could carry a human voice?

What if the sky itself could serve as a messenger?

In the quiet spaces of his family home in Italy, a young Guglielmo Marconi listened more to the air than to conversation. The year was 1894. The world was still tightly bound by copper wires and telegraph lines. Yet Marconi, not yet twenty, was drawn to an unseen realm that James Clerk Maxwell had predicted through mathematics and Heinrich Hertz had briefly proven with experiments. Many believed the space between transmitter and receiver was empty. Marconi sensed it was alive with possibility.

Without formal training in physics, he transformed his father's attic into a working laboratory. Surrounded by coils, transmitters, and carefully assembled circuits, he began sending wireless signals across short distances. From room to room. Then across open fields. Then between hills. While the world continued tapping messages through wired telegraph systems, Marconi quietly removed the wire altogether.

In 1896, he secured one of the world's earliest patents for wireless telegraphy in England. Two years later, his signals crossed the English Channel. Then, in 1901, standing on the rugged coast of Cornwall, he transmitted a faint but historic signal across the Atlantic Ocean to Newfoundland. Three short pulses. The letter "S" in Morse code. It became the first successful wireless transatlantic communication. At that moment, the planet grew smaller.

This discovery did not merely improve communication. It reshaped it. From Marconi's early transmissions emerged the foundations of radio, radar, satellite communication, and eventually the mobile phones that now rest in nearly every human palm. Yet in those early years, Marconi was not simply building devices. He was deciphering a new language of waves. He was learning how frequency could become meaning, and how silence itself could be taught to speak.

In 1909, Marconi was awarded the Nobel Prize in Physics, shared with Karl Ferdinand Braun. Yet his greatest legacy cannot be measured in medals. It exists in motion.

It travels with every call that crosses continents.

With every broadcast moving through space.

With every unseen message exchanged between distant minds.

Marconi not only invented a machine.

He opened the air itself and taught it how to carry thought.

Chapter 11
The Star Who Invented the Signal of the Future

In the golden age of black-and-white cinema, she dazzled the world. Her face was a portrait of perfection, her voice smooth, her presence magnetic. To audiences, she was a movie star. But behind the glamour, beyond the scripts and the lights, lived a mind tuned to equations, blueprints, and ideas far ahead of its time.

Hedy Lamarr
MGM studio publicity portrait for Comrade X, 1940

Her name was Hedy Lamarr.

Born Hedwig Eva Maria Kiesler in Vienna in 1914, she showed an early fascination with how things worked. While other girls played with dolls, Hedy took apart music boxes to study their inner mechanics. She was curious about motion, patterns, and systems. Yet her beauty was so striking that long before her scientific gifts were recognized, she was discovered by filmmakers.

Before she turned twenty, she was already a star. In Europe, and later in America, she was celebrated as the most beautiful woman in the world.

But beauty was never her greatest asset. Her mind was.

The world watched her on the screen. She, meanwhile, watched the world with an inventor's eye.

During World War II, while many celebrities focused on entertaining troops or selling war bonds, Hedy took a different path.

She partnered with composer George Antheil, not to create music, but to solve a military problem. Together, they confronted a challenge facing the Allied forces: enemy interference with torpedo guidance systems.

Her idea was bold, unconventional, and far ahead of its time. If a radio signal could rapidly change frequencies in a synchronized pattern, it would be nearly impossible for an enemy to jam it. The logic followed the same principle as a player piano roll, where shifting positions guides sound in precise sequence. This method later became known as frequency hopping.

It was not the kind of proposal anyone expected from a Hollywood actress. That, in many ways, is exactly why it mattered.

In 1941, Lamarr and Antheil filed a patent for their system. It offered a way to guide torpedoes securely and invisibly through constantly shifting radio channels.

The United States Navy declined to implement it. The technology, they said, was too complex for practical use at the time.

History, however, has a long memory.

Decades later, the core principle behind that forgotten patent became the foundation of spread-spectrum communication. Today, it underlies Wi Fi, Bluetooth, and GPS. The invisible signals that now move through phones, satellites, and networks across the planet trace their origins back to those wartime sketches.

In a world that once judged her mainly by appearance, Hedy Lamarr quietly helped shape the future of global communication.

She never made money from the invention. She rarely spoke about it. Fame stayed with her, but recognition for her scientific contribution arrived late. Only in the final years of her life did institutions begin to formally acknowledge what she had given the world. In 1997, she received the Electronic Frontier Foundation Pioneer Award. By that time, she had long withdrawn from public life and did not appear to accept it in person.

Yet the irony endures.

Hedy Lamarr, once dismissed as merely a beautiful face, helped make possible the very technologies that now connect the modern world.

Not every signal is visible.

Some are released quietly and change everything.

Chapter 12
Thought That Changed How We Navigate the World

Finding Our Way: From Stars to Satellites

For thousands of years, we relied on the sky. Early humans lifted their eyes upward, tracing constellations to understand where they stood in the vastness of the world. Sailors followed stars across dark oceans. Desert travelers read the angle of the sun. Explorers memorized rivers, ridges, and distant trees as anchors of direction.

But this ancient system was fragile. A wrong turn. A misread map. A clouded sky. Any mistake could mean being lost, sometimes forever.

Today, all it takes is a glance at a screen. In a crowded airport, a foreign city, or a dense forest, your phone can pinpoint your location within meters. It feels like magic. But it is not magic. It is mathematics, satellites, political rivalry, and a century-old theory that changed how we understand time. It is GPS.

The Spark: A Beeping Ball in the Sky

In October 1957, a strange sound drifted from space. A steady, metallic beep from Sputnik, the Soviet Union's first artificial satellite. As it circled Earth, scientists in the United States listened. At first, it was only a signal of technological rivalry. Humanity has entered the space age.

Then something deeper clicked. By studying changes in the satellite's radio frequency, known as the Doppler shift, physicists realized they could determine Sputnik's exact location. And a second question quietly followed. If we can locate a satellite from Earth, could satellites help locate us?

That simple reversal of logic planted the seed of GPS.

The United States Department of Defense soon began developing a navigation system for submarines. This work eventually evolved into the NAVSTAR GPS program.

Born from Cold War tension and driven by scientific ambition, GPS was originally built for military precision.

Then tragedy shifted its destiny. After a civilian aircraft was shot down for accidentally entering restricted airspace, the US government made GPS freely available to the public. By the early 2000s, what began as a guarded military tool had become a global utility.

How GPS Works: A Silent Symphony in the Sky

At any given moment, at least twenty-four satellites orbit Earth. Each one sends precise signals downward, quietly announcing its position and exact time. They act like invisible clocks in the sky.

When your device receives signals from four or more satellites, it measures how long each signal took to arrive. Because the speed of light is constant, those tiny delays translate into distance. Combine those distances, and your location emerges with remarkable accuracy.

But there is a complication. Time itself behaves differently for those satellites.

Einstein and the Bending of Time

Albert Einstein's theories once seemed abstract and distant from everyday life. Yet today, they guide your car, your phone, and even rescue missions.

According to relativity, time passes at different rates depending on speed and gravity. GPS satellites move at extreme velocity and operate far from Earth's gravitational pull. As a result, their onboard clocks drift slightly out of sync with clocks on the ground.

The difference is measured in nanoseconds. Yet without correcting for it, GPS errors would grow into kilometers within a single day. Einstein's equations do the quiet work of keeping the system honest.

Each time your phone guides you to a café, tracks a run, or helps locate a missing person, you are witnessing the practical power of a theory conceived more than a century ago.

Where GPS Touches Our Lives

GPS now operates in the background of modern existence. We rarely notice it, yet we depend on it constantly.

Travel and Navigation: Drivers, pilots, and sailors rely on GPS for safe and efficient movement.

Smartphones: Fitness tracking, mapping, social media check-ins, and weather monitoring all depend on satellite positioning.

Space Exploration: Space agencies use GPS-based systems to track spacecraft across immense distances.

Military and Security: Guided systems, troop coordination, and rescue missions depend on precise location data.

Agriculture and Environment: Farmers use GPS to guide machinery with exact efficiency. Scientists monitor land movement, oceans, and climate shifts using positioning data.

Emergency Response: From mountain rescues to disaster relief, GPS helps locate people when seconds matter.

The Road Ahead: Beyond Satellites

GPS continues to evolve. Scientists are now developing navigation systems that do not rely solely on satellites. Quantum positioning, underground navigation, underwater tracking, and even interplanetary guidance are all being explored.

What began as a faint Soviet beep in the cold silence of space, shaped by a German physicist's equations, refined through military urgency, now guides billions of people each day.

All of it grows from a single question that humanity has asked for as long as it has wandered:

Where am I?

And now, for the first time in our long history, we rarely have to be lost.

Chapter 13
Father Gave Him a Book, and He Gave a Book to the World

One afternoon, a high school boy came home from school to find his father absorbed in a quiet storm of thought. Papers, diagrams, and technical manuals were spread across the kitchen table. His father was preparing a speech and had drifted into a question that would not let him rest: could machines ever mirror the brain's extraordinary way of linking ideas?

Unlike computers, which store data in tidy rows and precise boxes, the human brain seems to move effortlessly between thoughts, forming connections that are fluid, intuitive, and sometimes mysterious.

Curious, the boy leaned in and listened as his father, a mathematician who had worked on one of the world's earliest computers, spoke aloud. "Computers store information in structures like lists or matrices," he said. "They are efficient, yes, but they cannot create the kinds of spontaneous associations that we do. Think of the smell of stale coffee, and how it suddenly carries you back to a memory. The brain weaves such links naturally, forming a vast web of knowledge that keeps evolving."

The boy nodded, quietly absorbing more than data. When the conversation ended, he walked off to do his homework. But the idea did not leave him.

What if computers could connect ideas the way the human mind does?

What if information itself could move freely from one thought to the next?

That boy was Tim Berners-Lee. His parents, Conway Berners-Lee and Mary Lee Woods, had worked on the Manchester Mark I, one of the first commercially available stored-program computers. But even before that evening conversation, another influence had already begun shaping Tim's thinking.

As a child, his father had given him a book titled Enquire Within Upon Everything. It was a Victorian-era reference book that covered an astonishing range of topics, from household advice to science and trivia. What fascinated Tim was not just the information inside, but the structure of the book itself. It invited the reader to search, to explore, to move from one question to another.

Its deeper message lingered: knowledge was not linear. It was interconnected.

That idea settled quietly into Tim's imagination.

Years later, he created a small software experiment and named it Enquire, in homage to that childhood book. The program was simple by modern standards, but it allowed users to organize and link information. A piece could connect to piece. Ideas could lead to ideas. It was a mirror, in digital form, of how human thought itself moves.

That modest experiment would later echo into something far larger.

The father's reflections on the brain, the childhood book, and the early Enquire software slowly converged into one of the defining ideas of the modern world: the World Wide Web.

Its concept was almost disarmingly simple. Information stored on different computers could be linked together. Anyone, anywhere, could follow those links, access knowledge, and add to it. In 1989, when Tim Berners-Lee first proposed the Web, it was a quiet idea in a quiet place. At first, the Web held only a few connected academic documents. Scientists shared notes. Researchers exchanged findings. It felt almost intimate.

But ideas, once released, rarely remain small.

Within a few years, the Web began to expand with astonishing speed. Static text welcomed images. Documents turned into living pages. Then came sound, video, and real-time interaction. The Web no longer reflected only human knowledge. It began to reflect human life itself.

Shopping, learning, debating, protesting, loving, mourning, celebrating, all found space inside this growing digital world.

Each visible slice of the Web now represents not just content, but human presence. Every website, every video, every message carries intent, emotion, labor, and identity.

Today, more than 1.2 billion websites exist, yet that number reveals only the surface. Beneath it lies the Deep Web, a vast domain of scientific data, medical archives, academic journals, private networks, and secured databases that search engines do not reach. And deeper still is the Dark Web, a shadowed space defined by anonymity, used for protection by some and for harm by others.

The scale of activity is difficult to fully grasp.

Over 500 million tweets are sent every day.

More than 347 billion emails move across the world every 24 hours.

Over 1 billion hours of video are watched on YouTube daily.

And this growth is no longer driven by humans alone. Machines now add their share. Artificial intelligence generates content. Bots populate platforms. Smart systems answer questions before we fully form them. The Web is no longer merely used by humanity. It now evolves alongside us.

What Tim Berners-Lee once envisioned as a web of knowledge has become something larger than he could reasonably have predicted. It began as an architecture of curiosity. It has grown into an engine of civilization.

What started as a childhood fascination with a book, a conversation across a table, and a simple software experiment has unfolded into one of the most consequential creations in human history. The World Wide Web, from its modest academic origins to its vast global presence today, stands as the greatest shared thought experiment humanity has ever conducted.

And like all living systems, it is still unfolding.

Chapter 14
The Dream That Rewired the World

Some dreams do not arrive in boardrooms or strategy sessions. They arrive quietly, in the middle of the night, uninvited and insistent. Larry Page once said, "When a really good dream comes up, grab it and run with it." For him, this was not a metaphor. During his graduate years at Stanford, he jolted awake from such a dream and reached for a pen before the idea could fade. What he wrote down was simple in language, but vast in consequence: to organize the world's information and make it universally accessible.

That fragile note in the dark would go on to ignite a technological revolution.

By the mid-nineteen nineties, the World Wide Web was no longer a curiosity. It had become a growing jungle of scattered pages, expanding faster than any system designed to organize it. Early directories, such as Yahoo, attempted to catalogue the web manually, much like a traditional library index. But human organization had limits. The scale of digital knowledge was already slipping beyond human hands.

It was within this rising chaos that Larry Page and his fellow graduate student Sergey Brin began experimenting with a research project they called Backrub. Unlike existing search engines that focused on direct keyword matches, their idea turned toward relationships. They paid attention to the invisible connections between websites. A hyperlink, they realized, was not just a technical bridge. It was a vote of trust. And votes could be measured.

Out of this insight emerged the PageRank algorithm. Instead of ranking pages by how frequently a word appeared, it ranked them by how many other pages linked to them and how authoritative those linking pages were. Relevance became relational. Authority became mathematical. This was more than an improvement in search. It was a new way of structuring knowledge itself.

They later renamed the project Google, a playful reference to the word "googol," which represents the number one followed by one hundred zeros. It was a quiet signal of their ambition to manage volumes of information that bordered on the unimaginable. What began inside a university lab soon escaped academic walls. The web was restless and disorganized. Google introduced clarity. It was fast, intuitive, and strikingly precise. Soon, people no longer only searched the internet. They Googled it.

In The Innovators, Walter Isaacson describes this moment as a turning point in the digital revolution. Page and Brin were not the first to dream of organizing human knowledge, but they transformed how that dream could function on a real scale. Their approach fused intuition with engineering discipline. In spirit, it echoed earlier visions such as Vannevar Bush's Memex and Ted Nelson's concept of hypertext, but this time the idea became operational, global, and immediate.

Google was not simply a tool. It reset expectations. Answers were no longer something you prepared for in advance. They arrived in seconds. Research that once required hours in libraries now unfolded in moments. Workflows narrowed. Attention compressed. Even human habits of memory and recall began to shift quietly beneath the surface. The collective human mind had found a digital extension.

And yet, this was only the first phase.

Information no longer wanted to remain bound to desks, cables, or fixed locations. A new transformation was already forming at the edges. With the rise of mobile technology and wireless connectivity, knowledge would soon begin to move with us, alongside us, and within constant reach.

The dream that began as a scribble in the dark was far from finished. It was preparing to leave the desk and enter the pocket.

Chapter 15
The Thought to Touch What You See

For centuries, we interacted with machines through layers of separation. We pushed buttons, turned knobs, and pressed keys. Our commands were always indirect. A mouse guided a cursor. A keyboard translated thought into text. Yet, deep within us, a quiet instinct persisted. A simple human question kept returning: why can I not just touch what I want?

That single desire, ordinary and intuitive, planted the seed for one of the greatest interface revolutions in human history.

The origins of this shift reach further back than many imagine. In the 1960s, a British engineer named E. A. Johnson was working at the Royal Radar Establishment. His aim was not consumer convenience, but efficiency in air traffic control. To manage flight data more intuitively, he developed a screen that could be touched directly. No knobs. No keyboard. Just finger to surface. At the time, it was a quiet experiment. In hindsight, it was the first visible step toward a future we now take for granted.

During the 1970s, that early idea gained a physical form. At Elographics, Dr. Samuel Hurst invented the resistive touchscreen. It worked by detecting pressure between two conductive layers. These screens were durable and precise. They soon appeared in ATM machines and industrial control panels. They worked reliably, but they demanded effort. You had to press. Interaction was still mechanical, not yet natural.

By the 1980s, another concept began to take shape. Researchers in Toronto, led by Dr. Tom Furness, explored capacitive touchscreen technology. Unlike resistive systems, this surface sensed the electrical properties of the human body. It did not require pressure. A gentle touch was enough. This was more than a technical improvement. It marked a deeper shift in how humans and machines would relate. The machine no longer needed force. It began to respond to presence.

Still, these systems remained confined to laboratories and specialized environments. They were expensive, experimental, and unfamiliar to the public. Then, in 2007, the waiting ended.

Apple introduced the iPhone.

For the first time, a capacitive multi-touch interface entered everyday life. People could swipe, pinch, zoom, and tap using only their fingers. No stylus. No physical keyboard. No instruction manual. The interaction felt immediate and instinctive. Within weeks, it became second nature. The iPhone did not merely introduce a new device. It erased the memory of the old way of using machines.

A barrier vanished.

Children navigated screens without explanation. Elderly users learned to browse photos and call loved ones with simple gestures. The interface did not demand learning. It adapted to human behavior.

And the change did not remain confined to phones. Tablets followed. Then, interactive kiosks, smart televisions, car dashboards, coffee machines, ticket counters, and medical devices. A flat sheet of glass became a gateway to work, memory, entertainment, and communication.

Touchscreens not only improve usability. They changed the philosophy of interaction. Before them, humans learned the logic of machines. After them, machines learned the logic of humans.

From here, the next transitions unfolded naturally. If machines could respond to touch, why not to voice? Why not move? Why not to faces? Voice recognition allowed us to speak commands aloud. Gesture control enabled interaction through motion. Facial recognition turned the human face into a living password.

We were no longer just operating machines. We were communicating with them.

In this new interface age, technology softened. It no longer felt distant or cold. The smartphone in your hand became more than a tool. It became a companion.

An extension of memory. An extension of intention.

This shift was not simply an innovation. It was evolution.

From pressing buttons to pressing thought into glass. From machines obeying structured commands to machines adjusting to human behavior.

And the transformation continues.

Flexible OLED displays now bend and fold. Haptic feedback introduces the illusion of physical texture, allowing virtual buttons to feel real. Researchers are experimenting with invisible interfaces, where gestures in open air replace physical contact. Tools such as Microsoft Kinect and Leap Motion hint at futures where screens may not even need to be touched.

Wearable technology brings this language of touch to the wrist. Smartwatches respond to taps and swipes on surfaces no larger than a coin. Control has become light, fast, and intimate.

All of this traces back to a very old human impulse.

Why can't I just touch what I want?

That thought did not disappear. It moved quietly across decades. It passed from one laboratory to another. From engineer to engineer. From idea to prototype. And one day, it arrived in your hand.

You swipe once.

And the world opens.

Chapter 16
When I Speak into My Phone: A Thought That Travels

Like most people, I speak into my phone every day. I talk. I listen. I share pieces of my life. The gesture feels ordinary now, almost invisible. Yet every so often, a quiet question returns and refuses to leave.

How does my voice, shaped only by breath and muscle, cross oceans and continents and arrive in another person's ear as though I were standing beside them?

There is no wire connecting us. No visible bridge. Still, the conversation flows without interruption. It feels like magic. But it is not.

It is thought wrapped in physics, carried by engineering, and guided by an invisible architecture of waves.

The First Breath of a Voice

Every call begins with something simple. It begins with me.

A breath leaves my lungs. My vocal cords vibrate. My tongue and lips shape the sound. In that instant, air becomes information. When I say "hello," I release tiny pressure waves into the surrounding space. They travel only a short distance before meeting the microphone inside my phone.

Inside that small opening sits a thin, not thicker than a strand of hair, sensitive enough to tremble at the faintest vibration. My voice touches it. The diaphragm moves. Motion becomes electricity. Sound is transformed into an electrical signal.

But electricity alone cannot travel cleanly across long distances. Not with the precision my voice requires. So the next transformation begins.

When Breath Becomes Data Inside the phone, an analog-to-digital converter begins its work. It does not understand words or meaning. It only measures change. Thousands of times each second, it samples the electrical signal and translates it into numbers. Each moment of sound becomes a sequence of digits. Ones and zeros. A digital fingerprint of my voice.

What began as breath has now become data. Not language anymore, but a mathematical description of vibration. This is the form machines understand best.

From My Phone to the World

Once my voice becomes numbers, it is no longer confined to my hand. It leaves the phone as radio waves, carried outward through the antenna. The signal reaches the nearest cell tower. These towers rest quietly among buildings, trees, and hills, relaying fragments of my voice across vast geographic distances.

Sometimes the journey is short. Sometimes it stretches across nations. When distance demands more, the signal enters fiber optic cables laid across the ocean floor. Light pulses through glass thinner than a strand of hair. Information moves faster than any physical courier ever could.

If the destination is even farther, the call may rise upward to a satellite orbiting the Earth. From there, the signal falls back down into another device waiting on the opposite side of the world.

All of that movement happens in less time than it takes to blink.

When My Voice Returns

At the destination, the process reverses.

Inside the receiving phone, a digital-to-analog converter reads the incoming numbers and reshapes them into electrical signals. The speaker diaphragm moves again. The air vibrates once more. And suddenly, my voice appears. Not as a copy. Not as an echo. But it sounds alive again.

Someone hears me say "hello."

WhatsApp, Wi Fi, and the Same Ancient Idea

Sometimes I do not call through traditional networks. I use internet-based systems such as VoIP. Here, my voice becomes small data packets. Each packet travels a different route through the internet, passing through servers, routers, and cables. At the destination, the fragments reassemble in the correct order.

The paths may differ. The technologies may evolve. But the essence never changes. Breath becomes data. Data becomes breath again.

Where It All Began

This journey began in 1876, when Alexander Graham Bell first turned sound into electricity. A simple wire. A transmitter. A whispered sentence. And suddenly, distance folded inward.

What followed was not just a sequence of inventions. It was a long reply to an ancient human question.

Can a machine listen?

Can it carry a voice beyond the reach of the body?

Can it return it intact?

The Echo of a Thought

Now, when I speak into my phone, I remember that my voice does not simply travel. It transforms. Thought becomes breath. Breath becomes vibration. Vibration becomes data. Data becomes light. Light crosses the world. And then the process unwinds until sound lives again.

What we casually call a phone is, in truth, the physical shape of a thought. Born in silence. Carried by science. Returned to life through sound.

Chapter 17
Social Media – Connecting Every Human Being Instantly

There was a time, not long ago, when a message traveled only as fast as a horse could ride or a ship could sail. Words folded into envelopes crossed mountains, deserts, and oceans in weeks, sometimes months. Those who waited measured time in uncertainty, staring toward the horizon, hoping.

Then came the first signals of acceleration.

The telegraph translated language into rhythmic taps that raced across wires. The telephone carried voices through metal veins. Email pulled conversation onto glowing screens. Each invention narrowed the distance between human beings. Yet none prepared us for what followed.

What arrived next was not built in factories. It was imagined in dorm rooms and garages. It was not driven by engines, but by connection. A simple and radical idea took hold: every human being on Earth could be linked, instantly.

What began as a digital whisper became a global chorus.

It started with the simplest human impulses. A photo of a newborn. A message to a friend across an ocean. A thought that wanted to be shared instead of forgotten. A desire not to feel alone.

Early platforms such as Friendster and MySpace explored this new frontier. Then, in 2004, inside a modest Harvard dorm room, a young student named Mark Zuckerberg launched Facebook. It was born not merely as a website, but as a window. A window into other lives, other thoughts, other worlds.

After that, the expansion was relentless.

Twitter compressed ideas into lightning-fast statements. Instagram transformed memories into curated images. LinkedIn reshaped the handshake into a digital introduction. WhatsApp and WeChat carried private conversations across continents.

Snapchat lets emotions appear and vanish in seconds. TikTok turned creativity into a global stage where strangers applauded from every corner of the world.

What made this transformation unstoppable was not the technology itself. It was the human instinct beneath it.

The hunger to belong.

They need to be seen.

The desire to speak without waiting for permission.

Social media became more than a tool. It became a mirror. And when humanity looked into it, it did not only see others. It saw itself.

In a single generation, social media reshaped how people learn, fall in love, argue, vote, shop, and dream. News no longer waited for printing presses. It erupted online. Movements such as the Arab Spring and MeToo did not rise only in the streets. They surged across screens and cities at once. A teenager in a remote village could sing into a phone and reach millions before sunset.

Time itself seemed to compress. A post shared at midnight could change a life by morning. A comment could ignite a protest. A video could shift global opinion in seconds.

Yet every acceleration carries its shadow.

Falsehoods began to spread faster than truth. Privacy became fragile. Mental health is strained under the pressure of comparison and curated perfection. Algorithms, once invisible, started shaping what people saw, what they believed, and even what they feared.

Humanity found comfort in connection, and chaos within it. People became visible and vulnerable. Voices grew louder, yet listening often grew weaker.

The original impulse was simple: connect people. Over time, it deepened into something far more complex. It became a system that links minds, emotions, and movements across time, space, and culture in real time.

It handed every individual a megaphone.

Every voice became a ripple.

Every ripple carried the potential to become a wave.

Where this leads remains uncertain. Will people one day gather in digital worlds as they once gathered in town halls? Will thoughts move directly from mind to mind without screens or keyboards? Will deeper unity emerge, or sharper division?

Social media is no longer just a platform. It has become a living network of human thought, moving faster than any previous force in history.

And within its orbit, humanity continues its search for meaning, belonging, and truth.

Chapter 18
The Thought That Went Digital

Some revolutions arrive with bloodshed. Others begin with blueprints. The digital revolution began with something quieter: a question in the human mind. *What if* thought could be transferred to metal?

ENIAC

ENIAC (1945) – First Electronic General-Purpose Computer, Built at the University of Pennsylvania for the U.S. Army.

What if logic could be wired? *What if* memory could exist outside the brain?

Long before screens lit our faces, humanity was already inching toward a new form of intelligence. Not organic, but coded. A world where memory would no longer be bound to the limits of the mind, and communication would no longer be restrained by distance.

ENIAC and the Birth of Machine Thinking

The earliest computers were not born from comfort or convenience. They were born from urgency. War demanded faster calculation, and massive machines answered that call.

The ENIAC was one of the first of its kind. A colossal assembly of wires, vacuum tubes, and blinking lights, it filled entire rooms and consumed enormous power. Its task was precise: to calculate missile trajectories at speeds no group of mathematicians could match. These machines did not belong to individuals. They were housed in institutions, guarded, respected, and rare.

They could compute, but they could not yet belong to everyday life.

Then came the transistor. That single invention reshaped everything. Giants became smaller. Machines became faster. Costs dropped. With every generation, computers shrank in size while expanding in power. A new idea quietly took hold. *What if* every home could have one?

In garages, dorm rooms, and small workshops, young innovators began shaping a future no government laboratory had fully imagined. Within a decade, a plastic box on a desk became more powerful than entire rooms of machines from the past. The computer was no longer distant. It was personal.

Yet computing alone was not enough. Processing information was only the beginning. The deeper transformation came with connection.

The Internet and the Web of Minds

In government laboratories, another quiet revolution was unfolding. The goal was simple in design, but radical in consequence: connect distant computers so they could share information. ARPANET became the first visible thread.

Still, it took the vision of Tim Berners Lee to give the internet its human shape. He imagined a web that would connect not just machines, but people, pages, and ideas. The World Wide Web turned abstract networks into a shared global space. And once that door opened, it could not be closed.

The 1990s surged with digital momentum. Email replaced letters. Search engines began replacing encyclopedias. Dial-up tones announced the arrival of a new era in ordinary homes. Shopping crossed from streets to screens. Portfolios moved online. Conversations transformed into chat windows. Information, once slow and distant, became instant and expected.

But something more human was happening beneath the technology.

The Rise of Social Identity Online

People wanted more than access to data. They wanted presence. They wanted to be seen and heard. The internet slowly gave birth to platforms where identity could be shaped, opinions magnified, and emotions exchanged in real time.

Social media did not emerge simply as a tool. It emerged as a phenomenon. From MySpace to Facebook, from Twitter to TikTok, the digital world became a second public square. It connected continents, yes, but it also began to rewire attention, memory, and behavior.

Scrolling started to replace deep reading. Sharing often replaced reflection. Memes traveled faster than essays. A single post could ignite a movement or inflame a crowd. Thought, once confined to paper and voice, now moves across the planet in seconds.

The power was unprecedented. And so were the consequences.

Artificial Intelligence and the Machine That Learns

Then came the next threshold. Could machines go beyond storage and calculation? Could they begin to think in their own way?

Artificial intelligence did not arrive with spectacle. It entered quietly. Into search engines that predicted our queries. Into voice assistants that answered our questions. Into cameras that recognized faces. Into vehicles that learned how to drive.

AI does not simply imitate intelligence. It learns. It adjusts. It improves. With every dataset, it grows closer to systems that do not just process thought, but generate it.

This shift may one day rival the invention of language itself.

The Shadow of the Digital Age

With each breakthrough, a shadow followed.

Data theft. Digital addiction. Misinformation. Surveillance capitalism. The very tools designed to expand freedom began demanding attention, personal information, and time. Privacy became negotiable. Truth grew increasingly filtered through algorithms. A widening gap separated those with digital access from those without it.

The digital world is empowered. It was also extracted.

The Fastest Acceleration in Human History

Still, the digital revolution remains one of the most rapid transformations the human species has ever experienced. It flattened hierarchies. It democratized knowledge. It created industries built entirely from code. It allowed a single idea, shared at the right moment, to reshape economies, politics, and culture itself.

And it all began with a thought. A quiet question. A subtle spark.

What if thinking could become digital?

Chapter 19
When Machines Found Their Voice: A Whisper,
A Wave, A Wonder

It did not begin with a revelation. It began with a question.

A quiet voice rose from a small box on a table:

"How can I help you today?"

You hesitated. Then you answered.

The machine listened. I understood. It replied.

Not by magic.

Not by accident.

But through the long, patient unfolding of human thought.

For the first time in history, sound crossed a boundary that had separated the living from the lifeless. A machine did not merely react. It listened. And then, it spoke.

From Sound to Meaning

To speak is human.

To listen is intelligent.

To speak back is where the machine begins to feel alive.

When you speak, your voice does not travel like words. It moves like a vibration. Waves of pressure roll through the air and strike a microphone like ripples across water. The microphone converts these ripples into electrical signals. But machines do not understand vibration. They understand numbers.

Your voice is broken into fragments, measured, digitized, and translated into data.

This is where Automatic Speech Recognition, or ASR, enters. It is the first true interpreter between sound and sense. ASR does not simply hear noise. It searches for patterns. It compares frequencies. It tests possibilities. It reconstructs syllables. Then words. Then meaning.

In less than a second, breath becomes data. Sound becomes language.

What once required a human ear now happens inside a processor.

The Long Road to Understanding

The earliest machines that attempted to recognize speech were fragile listeners. In the 1950s and 60s, they could detect only a few digits spoken by a single trained voice. They did not really understand. They matched patterns the way a lock matches a key.

Progress was slow. For decades, machines struggled with accents, speed, background noise, and emotion. Human speech is chaotic. We pause, interrupt ourselves, whisper, shout, and change direction mid-sentence. Teaching order to interpret disorder took generations of research.

Real transformation arrived with machine learning and deep neural networks. Instead of following fixed rules, machines began to learn the way humans do. By exposure. By experience. By correction. Millions of voices trained them. Billions of words shaped them.

What followed was not a leap forward, but a quiet crossing of a threshold.

Machines stopped merely detecting sound.

They began to grasp intent.

When the Machine Speaks Back

Understanding is only half of the conversation. Response is the other half.

Once a machine knows what you said, it must decide what to say next. This is where the anatomy of machine speech unfolds in layers:

ASR converts your voice into text.

Natural Language Understanding interprets the meaning.

Natural Language Generation decides on a response.

Text-to-Speech gives that response a voice.

In the earliest systems, machines assembled speech from fragments of recorded human words. The result was functional but flat. Clear, yet cold. The voice spoke, but it did not breathe.

Then neural speech synthesis arrived.

With systems such as WaveNet, machines no longer replay sound. They generated it. Each syllable was shaped, not selected. Rhythm emerged. Intonation softened. Pause and emphasis entered the voice.

For the first time, machines not only speak. They performed speeches.

The difference was subtle, but unmistakable. The voice no longer sounded assembled. It sounded formed.

Where Thought Meets Daily Life

This technology does not live in laboratories alone. It speaks from kitchens, classrooms, hospitals, and fields.

A blind student uses voice guidance to move through a crowded school.

A farmer listens to spoken weather alerts before planting a single seed.

A grandmother asks a smart speaker for an old family recipe, just to hear it spoken aloud again.

These machines do not speak with their own voice.

They speak with us.

Every accent, every hesitation, every question ever asked has trained them. They reflect not a foreign intelligence, but a collective human one.

What we hear when a machine speaks is not its mind. It is the echo of, ours.

Still Becoming

The question that once seemed impossible was simple:

Can a machine speak?

That question has already been answered.

A deeper one now takes its place:

What will it say next?

Perhaps machines will one day carry a full emotional range in their voices. Perhaps they will learn persuasion, comfort, and empathy not as programs, but as patterns of response refined over time. Or perhaps limits will remain. The future is not certain. It is a probability.

What is clear is this: every conversation we have with a machine is a mirror. It reflects our needs, our habits, our patience, our urgency.

In the beginning, we gave machines ears.

Then we gave them language.

Now, we have given them a voice.

The thought behind that voice, however, still belongs to us.

Chapter 20
Blockchain – A Ledger of Thought

Can trust exist without a middleman? Can value move across the world without a bank, a government, or even a verified name attached to it?

For most of human history, the answer would have been a firm no. Trust needed a face, a seal, a crown, or a stamped signature. Empires were built on ledgers guarded by clerks. Banks rose as temples of verification. Nations stood as guarantors of currency. And then, in the shadow of the 2008 financial collapse, when faith in these towering institutions cracked in full public view, a quiet document appeared on the internet.

It was not announced at the ceremony. There was no press conference, no identity, no photograph. Just a whitepaper shared on a cryptography mailing list. The name beneath it was unfamiliar and remains so to this day: Satoshi Nakamoto. Whether a single person or a collective mind, no one knows with certainty. What the document proposed, however, was clear and radical. A peer-to-peer electronic cash system. Money without a central authority. Trust without a gatekeeper.

The invention was called Bitcoin. But beneath the currency was something far more enduring: blockchain.

A Ledger Without a Gatekeeper

Long before we trusted algorithms, we trusted people. Kings validated land. Priests recorded births. Banks confirmed wealth. Every transaction in civilization has depended on a central witness. Someone said, "This happened. This belongs to you. This is real."

Blockchain asks a startling question: *What if* the record itself became the witness?

At its core, blockchain is a digital ledger that anyone can inspect and no single authority can control. Imagine a book where every page contains a batch of verified transactions. Once the page is full, it is sealed by cryptography and permanently linked to the page before it. No erasing. No editing. No rewriting of history. Each new page strengthens the entire chain behind it.

This ledger is not stored in one office or under one government. It lives at once on thousands of computers across the world. These computers, called nodes, constantly check one another. If someone tries to alter a record, the network rejects it instantly. Truth is not enforced by power, but by agreement.

No headquarters. No master key. Only consensus.

For the first time in history, humans created a system where trust emerges not from authority, but from mathematics.

The Currency of Thought

Bitcoin was the first expression of this new logic. It allowed people to send value directly from one person to another across borders, without a bank standing in the middle. For some, it felt liberating. For others, unsettling. If money could exist without a state, then what else could be redesigned?

Soon, it became clear that blockchain was larger than Bitcoin itself.

Ethereum expanded the idea beyond money. It introduced smart contracts, self-executing agreements that carried out their terms automatically once conditions were met. No lawyer was needed. No courtroom waited in the background. Code replaced the referee.

From there, innovation moved quickly. Litecoin aimed for faster payments. Ripple focused on international transfers. Stablecoins attempted to tame volatility by tying digital currency to real-world assets. Thousands of alternative tokens followed, some thoughtful, some reckless, all testing the boundaries of what value could become.

Then came new waves. Initial Coin Offerings promised crowd-powered investment. NFTs redefined digital ownership. DeFi, or decentralized finance, has built entire financial systems without banks. Lending, borrowing, and trading all unfolded through software alone. Art moved without galleries. Contracts moved without courts. Ownership moved without paper.

It felt, to many, like standing at the edge of a financial ocean whose depth no one yet understood.

Power and Paradox

Every great technological leap arrives carrying both promise and tension. Blockchain is no exception.

It offers transparency, yet it can also enable anonymous crime. It distributes power, yet it demands enormous energy to sustain certain networks. It provides financial tools to the unbanked, yet it confuses regulators tasked with protecting stability. It resists central control, yet governments now race to integrate versions of it.

Banks fear its disruption, even as they quietly build their own blockchain systems. Environmentalists criticize its energy use, especially in early proof-of-work systems. Technologists continue to refine it, seeking efficiency, speed, and sustainability.

At the same time, new visions emerge. Central banks explore digital currencies. Decentralized identity systems promise proof without exposure. Interoperable chains aim to allow blockchains to speak to one another, much like once-divided cities learned to trade across walls.

The debate is not settled. It is evolving, minute by minute, as code updates and markets move.

A Philosophical Shift in Silence

Beneath all the technical language, blockchain represents a profound shift in how humans think about trust. For centuries, value was declared. A king stamped a coin. A nation printed a note. A bank certified a balance.

With blockchain, value is verified not by decree, but by math and collective agreement.

It quietly challenges the idea that governments alone should define money. It hints, perhaps, at a future where individuals hold greater custodianship over their wealth and identity. It also raises difficult questions about responsibility, regulation, and the risks of unchecked systems.

A teenage artist in Manila can now sell digital art to a collector in Paris without ever sharing a language. Two businesses on opposite sides of the planet can exchange value in minutes without opening bank accounts in each other's countries. These exchanges are not acts of trust in people. They are acts of trust in code.

Financial inclusion, once framed as charity or policy, begins to resemble protocol.

A Single Thought That Shook the System

Satoshi Nakamoto did not build a bank. Did not seek recognition. Did not claim power. Instead, a single idea was released into the world and left to grow on its own. A question, more than a system. Can trust be decentralized?

In the years since, fortunes have been made and lost. Markets have swung wildly. Scandals have erupted. Innovations have multiplied. Yet beneath all of this noise, the deeper shift remains.

Blockchain taught humanity that trust can be engineered, not just declared.

Whether Bitcoin becomes the future of money or a stepping stone toward something else, the wave it unleashed will not recede quietly. It has already rewritten how people think about ownership, value, and agreement in a digital age.

In this unfolding chapter of human thought, we are not just watching a financial experiment. We are witnessing a subtle rewiring of how humans agree with one another about what is real, what is owned, and what is worth believing in.

Chapter 21
3D Printing – Creating Reality Layer by Layer

For most of human history, creation meant subtraction. We carved stone to reveal a shape, shaved wood to form a tool, and hammered metal until it obeyed our intentions. Making something was heavy, loud, and bound by the limits of hand and material. Yet quietly, in the late twentieth century, a different idea emerged. *What if* objects could be built the way we print words on paper? *What if* creation could rise upward, layer by layer?

That thought began as a technical curiosity. In 1981, Dr. Hideo Kodama explored a method of solidifying liquid resin with ultraviolet light. Three years later, Chuck Hull introduced stereolithography and with it the STL file format that would become the universal language of early 3D printing. What started as an experiment in laboratories would soon ripple into factories, hospitals, and eventually the homes of ordinary people.

The principle behind it is simple but profound. A digital design is sliced into extremely thin layers. A printer then deposits material, plastic, resin, metal, or even living cells one layer at a time. The object grows from the ground upward, as if thought is slowly gaining weight and settling into the world. Code becomes form. Imagination becomes something you can hold.

As patents expired, the technology escaped the walls of corporations. A child can now print a toy at home. A hobbyist can recreate a broken component without calling a manufacturer. An inventor can test a prototype the same afternoon the idea appears in their mind. Creation has moved from distant factories to the tips of our fingers.

Entire industries feel the shift.

Medicine. Doctors print custom prosthetics shaped exactly for a patient's body. Researchers are building early models of living tissues. Scientists have printed miniature hearts using real human cells. These are only early steps, yet they hint at a future where an organ can be printed for a single person, perfectly matched and ready for transplant.

Aerospace. NASA prints spacecraft components with geometric precision that traditional tools cannot achieve. In preparation for long missions, they experiment with printing both tools and food, reducing the need to carry large inventories into space.

Construction. Printers the size of small cranes now build entire houses in under a day. Concrete becomes a continuous line, rising into walls that are both strong and efficient.

Fashion and jewelry. Designers create intricate shapes that once defied conventional manufacturing. Weightless sculptures, complex lattices, and finely detailed ornaments move from imagination to reality with astonishing ease.

Each of these examples points to a larger shift. Manufacturing is moving from mass production to mass personalization. Instead of producing millions of identical objects, we create what is needed, when it is needed, shaped for one individual. Warehouses shrink. Waste falls. The supply chain becomes lighter and more intelligent.

And beneath all this progress lies a deeper truth about what it means to create. When I look at the evolution of 3D printing, I see more than machines and materials. I see a reaffirmation of human curiosity. A desire to shape the world not only through labor or force, but through understanding. A thought that grows quietly, layer upon layer, until it becomes something real.

The idea of printing reality once belonged to science fiction. Today it sits on desks, hums in studios, and saves lives in hospitals. From the first glimmer in a researcher's mind to the printed heart beating softly in a laboratory dish, the journey has taken only a few decades. Yet the impact reaches far into the future.

A thought rose from code. A shape emerged from light. And creation, once defined by chisels and hammers, now unfolds with the gentle precision of layers forming one atop the other.

This is the power of the human mind: to imagine, to refine, and now, to print what once could only be dreamed.

Chapter 22
The Thought of Internet of Things (IoT) – When Things Started Talking

There was a time when the world around us was quiet in a different way. Objects existed only to serve. They worked, they waited, and they forgot. A refrigerator cooled without knowing what it held. A watch kept time without remembering a single moment. A door locked with mechanical certainty and asked for nothing in return. The physical world performed its duties in silence.

Then a simple question entered human thought, gently but persistently. *What if* things could speak? *What if* the inanimate world could listen, respond, and share what it sensed?

This quiet curiosity gave birth to one of the most transformative ideas of the modern age: the Internet of Things (IoT).

The earliest signal of this shift did not come from a polished laboratory or a global technology firm. It came from a soda machine. In 1982, researchers at Carnegie Mellon University connected a Coca-Cola vending machine to the internet. The machine could report whether drinks were available and whether they were cold. It sounds modest now, almost playful. Yet it marked a turning point. For the first time, an ordinary object shared information across distances without human intervention. A machine, in its own limited way, had spoken.

Years later, in 1999, British technologist Kevin Ashton gave this growing idea a name. While working with Procter and Gamble, he imagined a system where everyday objects could be identified, tracked, and understood through data. His focus was practical. He wanted to make supply chains smarter and more efficient. But behind that practical goal was a larger vision. A world where physical objects could communicate what they sensed, where the silent materials of daily life could join the digital conversation.

If people could speak across the internet, why not the tools they built? If machines could measure their environment, why should that knowledge remain locked inside them?

At first, this shift moved quietly through homes. Thermostats learned daily routines and adjusted room temperatures without being asked. Refrigerators began keeping track of what was inside. Voice assistants turned spoken words into actions. Lights responded to movement. Doors unlocked at a touch. What once required constant attention now flowed in the background, almost unnoticed.

Healthcare soon felt the impact. Wristbands and smart watches began to track heartbeats, steps, sleep patterns, and oxygen levels. Subtle changes in the body could be detected early, sometimes before a person felt anything was wrong. Hospitals use connected devices to monitor patients continuously instead of relying on periodic checks. It did not replace doctors, but it gave them better sight.

Factories shifted as well. Machines learned to recognize early signs of wear. Repairs could happen before a breakdown halted production. Farms became more attentive to the land. Sensors measured moisture in the soil and guided irrigation only where it was needed. Water was saved. Crops grew with greater precision. Cities followed. Traffic lights are adjusted in real time. Power grids balanced demand more smoothly. Waste systems signaled when they were full.

The Internet of Things did not simply make life more convenient. It changed the rhythm of decision-making. The world began to anticipate rather than react. A flat tire no longer had to arrive as a roadside surprise. Sensors could warn before the pressure dropped too far. Electricity no longer flowed blindly. Smart grids are adjusted to usage in milliseconds. Objects were no longer passive. They had become participants.

Yet every new form of connection carries a shadow. With communication came exposure. The same data that was protected was also revealed. Who owns the heartbeat recorded by a watch? Who controls the patterns gathered from a home? The connected world raised questions humanity had never been forced to answer at this scale. Privacy, security, and responsibility became part of everyday life. Convenience demanded a new kind of ethical thinking.

The story of the Internet of Things did not stop with humans speaking to machines. It moved further into a world where machines speak to each other. A car signals a garage before it arrives. An alarm clock triggers a coffee maker. A home becomes a network of small digital conversations happening without awareness. This quiet exchange of signals formed a new kind of ecosystem.

Cities began to behave like living systems. Hospitals used predictive models to detect risk earlier. Farms produced more with fewer resources. Forests and oceans gained silent guardians through sensors that tracked fragile balance. The natural world, long observed by human instruments, was now being continuously listened to.

As I reflect on this shift, what strikes me most is not the speed of the machines. It is the speed of our trust. In just a few decades, we have invited invisible systems into the most private spaces of daily life. We allowed them to listen, to learn, and to act on our behalf. Some days this feels like liberation. Other days, it feels like surrender. Perhaps it is a little of both.

From silence to subtle exchange, from isolation to constant connection, the Internet of Things reshaped how the physical world meets the digital one. It is not merely a technology. It is thought to be made visible in motion. It is a curiosity given structure and instructions. For the first time in human history, the environment around us responds.

Objects no longer wait blindly for command. They respond. They adapt. They learn.

And in their quiet communication, we hear one of humanity's oldest questions again, returning in modern form. *What if* the world itself could think with us?

Chapter 23
Deep Learning – Going Beyond Human Thinking

For most of our technological past, we taught machines a single mode of existence. They followed our instructions, row by row, obedient and predictable. They processed information quickly, yet without imagination. Their world was a landscape of commands, not possibilities.

Then a simple but radical question surfaced: *What if* a machine could learn from experience instead of relying only on human rules? *What if* it could improve by recognizing patterns, much like the human brain does every moment?

This question became the seed of machine learning. And from that seed, another idea emerged, one that searched even deeper. Could a machine learn in layers, forming its own internal understanding of the world?

This thought opened the door to deep learning, a field where machines do more than compute. They begin to perceive. They begin to connect. They begin to construct meaning from complexity. In a subtle way, they step closer to the rhythms of human thought.

The Early Dreamers

In the 1980s, a small group of researchers believed that intelligence might be built from networks that resembled the brain. Geoffrey Hinton, Yann LeCun, and Yoshua Bengio explored the idea that artificial neurons could pass signals, strengthen connections, and gradually learn. Their work mirrored the living networks inside our skulls, where each memory, sensation, or idea is carried by the delicate cooperation of neurons.

Their designs were elegant, even visionary, but the world was not ready. Computers lacked the power. The data was limited. Their models were often dismissed as slow, impractical, or overly idealistic. So the idea slipped into the background, not defeated but waiting. Many meaningful ideas in science follow this pattern: a quiet period, a slow gathering of energy, and then an unexpected return when the world finally catches up.

The Silent Return

By the 2010s, the world had changed. We suddenly had the computational force to run complex neural networks. Graphics processing units became the engines of this new age, capable of performing millions of simultaneous calculations. The internet, overflowing with images, text, and sound, provided the raw material for learning. Data was no longer scarce; it was abundant.

This convergence brought deep learning back to life. Ideas that once felt premature became essential.

When Machines Began to Perceive

Deep learning systems learned to see. A convolutional neural network could identify edges, then curves, then entire faces, without ever being told what a face was. It discovered meaning by examining millions of examples, slowly building an understanding that rose from simple patterns to complex forms.

Machines learned to hear, too. They interpreted human voices, understood accents, and recognized emotion in tone. They learned to speak, translating languages, summarizing paragraphs, and generating answers that felt conversational.

Deep learning shifted the nature of artificial intelligence. Traditional algorithms required humans to define rules. These networks discovered their own rules, often surprising their creators. It was as if the machine opened its digital eyes and began to form its own quiet interpretations of the world.

Dreams Becoming Real

The impact quickly moved from theory to life.

Self-driving cars learned to read the road in motion, recognizing lanes, signs, and patterns in human behavior. Medical AI systems detected diseases inside scans long before doctors could see them. Voice assistants transformed machines from cold tools into companions that could hold simple conversations.

And then came moments that startled even the experts.

In 2016, AlphaGo defeated the world champion in Go, a board game with more possible moves than atoms in the universe. It was a turning point. Years later, AlphaFold solved one of biology's greatest puzzles by predicting the folded shapes of proteins. Scientists had worked on this challenge for decades. Deep learning completed it in months.

The Black Box Problem

But brilliance carried a tension. Deep learning models became powerful, yet their internal reasoning remained hidden. They could produce accurate results, and still, researchers often could not explain the step-by-step logic behind the decisions. Their minds were illuminated in outcome but shadowed in process.

This raised questions that went far beyond engineering.

If a machine can learn in ways we cannot fully understand, what does that reveal about the limits of our own understanding? Are we witnessing a reflection of ourselves, or are we building something fundamentally different?

These questions are not warnings. They are reminders that every new idea reshapes the way we see ourselves.

A Mirror to the Human Mind

Deep learning is not just a technological achievement. It is an insight into the nature of intelligence. It pushes us to consider what understanding truly means. It urges us to examine creativity, intuition, and awareness.

Perhaps intelligence is not tied to biological form. Perhaps it is a quality that can emerge wherever patterns learn from experience, adapt across time, and search for meaning. What began as a simple attempt to mimic the brain has grown into something far more revealing. In teaching machines to learn deeply, we have been reminded of the depth of our own minds. The boundary between human thought and machine thought no longer feels like a wall. It feels like a frontier. And like all frontiers in human history, it invites curiosity, caution, and wonder.

Chapter 24
Quantum Computing – The Next Frontier of Technology

There are moments in human history when thought pushes against the limits of what seems possible, and a new horizon appears. Quantum computing is one of those moments. It began as a question, not a machine. *What if* we could build computers that follow the rules of nature at its smallest scale rather than the logic of human engineering? The idea felt almost unsettling, yet it opened a doorway into a world that behaves nothing like the one we can see.

At the beginning of the twentieth century, physics uncovered a reality that did not obey classical intuition. Max Planck spoke of energy arriving in tiny packets. Einstein followed light across space and time, discovering that it could behave like both a wave and a particle. Niels Bohr watched electrons move not along predictable paths, but within clouds of probability. The universe, once understood as a tidy mechanical system, revealed itself as flexible, uncertain, and full of possibility. An electron could be in more than one position at once. A photon could behave in contradictory ways depending on who observed it. Reality seemed to interact with the mind that studied it, a conversation rather than a fixed script.

To understand why quantum computing is so radical, imagine a coin spinning in the air. While it rotates, it is not simply heads or tails but a blend of both possibilities. This is a superposition. Now imagine two coins connected in such a way that the moment one lands, the other reflects the result instantly, even if they are separated by a great distance. This is entanglement. These are not metaphors. They are the ways nature truly behaves, and from them a new form of computation emerges.

Classical computers rely on bits that hold clear values of zero or one. Quantum computers use qubits. A qubit can be zero, one, or both at the same time while it remains undisturbed. When qubits become entangled, they create a network of linked possibilities. A single instruction does not move through one pathway at a time. It ripples through many, exploring multiple outcomes together. It is a form of thought unfolding in parallel, not sequentially.

In the early 1980s, Richard Feynman questioned why we were trying to simulate a quantum universe with classical machines that do not follow quantum rules. It felt like trying to paint a rainbow using only black and white. If nature works through quantum behavior, perhaps our computers should as well. David Deutsch then proposed the idea of a universal quantum computer, one capable of processing many realities at once. In 1994, Peter Shor demonstrated that a quantum computer could break mathematical codes thought to be unbreakable. A few years later, Lov Grover introduced a quantum method to search data far faster than classical algorithms. These breakthroughs did more than solve problems. They showed that quantum computation was not a philosophical idea but a practical possibility.

Yet nature protects her deepest behaviors with delicate barriers. Qubits are extremely fragile. A slight change in temperature or vibration can erase their quantum state. Researchers had to invent new methods to protect them, correcting errors that appear before they can be detected, and designing environments where qubits could hold their uncertainty long enough to compute. Progress was slow for decades. Then, in 2019, something shifted.

Google announced that its Sycamore processor had performed a computation in minutes that would take a classical supercomputer thousands of years. The task itself was not meaningful outside the experiment, but the accomplishment was symbolic. A quantum machine had finally done something no classical machine could match. Since then, researchers at IBM, Microsoft, IonQ, Rigetti, and other institutions have been pursuing the next steps: qubits that are more stable, systems that can scale, and techniques that can carry quantum behavior from theory into real-world applications.

The potential applications are immense. Quantum computers could reshape cryptography, making current systems obsolete and requiring entirely new methods of security. They may allow scientists to model molecules with extreme precision, opening pathways to new medicines. They could design materials by manipulating atoms directly, rather than relying on trial and error. Artificial intelligence systems might learn faster and understand patterns too complex for classical machines. Quantum computing could illuminate processes ranging from protein folding within our bodies to the birth of stars across the universe.

Yet the greatest impact may be philosophical. Quantum computing challenges the idea that certainty is the foundation of progress. Instead, it reminds us that exploration begins in uncertainty. Much of human advancement arises from considering possibilities before choosing a direction, just as a qubit holds multiple potential states before observation. It is a quiet but powerful message. Curiosity, not certainty, is what moves us forward.

As we stand at the edge of the quantum frontier, it becomes clear that this technology is not only a scientific achievement. It is a shift in how we think. It allows us to imagine a world where computation mirrors the strange and beautiful rules of reality itself. And in doing so, it invites us to look inward. The human mind, with all its questions, intuitions, and flashes of insight, has always been the original system capable of holding many possibilities at once. Quantum computing is an extension of that capacity, a reflection of our desire to understand a universe that has never been as simple as it appeared.

The journey is only beginning, but its promise is profound. In reaching for quantum machines, we are learning something about the universe and something about ourselves. We are discovering that progress often begins with a question that seems too ambitious, too uncertain, or too strange. Yet those questions are where the future waits.

Chapter 25
ChatGPT – Talking to Machines

Not long ago, when we spoke to machines, they answered with silence wrapped in math. Commands went in. The results came out. The exchange was precise, fast, and emotionally empty. The machine executed, but it did not converse. It obeyed, but it did not seem to listen.

Then something shifted.

Today, you might be sitting alone in a quiet room with nothing more than a laptop in front of you, and the machine does more than calculate. It answers in full sentences. It explains. It asks clarifying questions. At times, it even appears to reflect. You type, and the response feels less like an output and more like a reply.

This is no longer the language of science fiction.

This is ChatGPT.

When I first watched a machine respond with fluid language, I felt a familiar sensation from earlier chapters of this book. The same feeling that must have stirred when the first symbols were carved into clay, or when the first voice traveled across a telephone wire. Another boundary had quietly moved.

The Question That Changed Everything

The idea behind ChatGPT began with a daring question. *What if* machines could talk not in rigid commands, but in natural conversation?

That question took a concrete shape in 2017, when a team of researchers at Google introduced a breakthrough called the Transformer. This neural architecture could do something fundamentally new. It could pay attention.

Not in the human sense of awareness, but in a mathematical one. It could scan language, look backward and forward within a sentence, and trace the subtle web of relationships between words. Meaning was no longer locked into straight lines. It became a network.

For the first time, a machine could follow context with a depth that had previously been impossible.

OpenAI carried this idea forward and expanded it into something larger. From that work emerged GPT, short for Generative Pre-trained Transformer. Unlike earlier programs that relied on fixed rules and hand-built dictionaries, GPT was designed to learn from exposure.

It was not told what language was. It discovered language by reading it.

How ChatGPT Learns to Speak

The learning journey of ChatGPT unfolds in stages, each one carefully designed.

First comes pre-training. During this phase, the model studies vast amounts of text. It absorbs the rhythm of language, the way sentences rise and fall, the patterns that give meaning its shape. It does not memorize in the human sense, but it learns probabilities. What word is likely to follow another. What structure tends to form meaning.

Then comes fine -tuning. Here, human guides gently shape its responses toward usefulness and safety. The goal is not just accuracy, but clarity and helpfulness.

Finally, through reinforcement learning from human feedback, the model learns how people react to its answers. It adjusts. It improves. It becomes more aligned with how humans communicate.

The result is something quietly astonishing.

Ask it to write a sonnet about Newton's laws, and it can do so. Ask it to summarize a complex scientific paper, and it can translate dense research into clear language. Ask it for a story set on Mars, and it will imagine one instantly.

Behind every response lies not intuition or feeling, but an immense statistical understanding of how language works.

A New Doorway to Knowledge

Beyond novelty, ChatGPT represents something larger. It marks a turning point in how knowledge is accessed.

A student without a skilled teacher. A farmer seeking better methods. A business owner refining an idea. A grandmother searching for a forgotten poem. All now approach the same doorway. The distance between expert knowledge and everyday life has narrowed.

This is not a small change.

In earlier centuries, knowledge moved slowly. It traveled on foot, by ship, by printed page. Access depended on geography, money, and privilege. Today, a question typed into a screen can summon centuries of accumulated information in seconds.

For the first time in human history, conversation itself has become a gateway to global knowledge.

That reality still feels almost unreal to me when I pause and reflect on it.

The Mirror of Human Intelligence

Every great tool eventually turns into a mirror. The wheel showed us how movement could be extended. The telescope showed us how vision could stretch across space. ChatGPT reflects something even more intimate.

It reflects thought.

If a machine can mimic reasoning, where does human thought truly begin? If a system can produce emotional language, what does that say about the structure of emotion itself? If it can persuade, where do we draw the line between assistance and influence?

ChatGPT does not feel. It does not desire. It does not intend to. Yet it mirrors the patterns of human expression with unsettling precision. In doing so, it forces us to look more closely at our own minds.

Perhaps that is one of its quietest powers.

From Cave Walls to Conversations with Code

From the first cave paintings to hieroglyphs, from the printing press to the telephone, every era has expanded how humans speak and how far their voices can travel.

ChatGPT is the latest ripple in that long wave.

We once used rocks to mark meaning. We later used ink. Then electricity. Now we use algorithms shaped by billions of human sentences. The medium keeps changing, but the impulse remains the same. To express. To ask. To understand.

Talking to machines is no longer a novelty. It is becoming part of ordinary life.

Unsettled Questions and Future Paths

With every leap of thought comes unease. The rise of conversational machines raises difficult questions that we are only beginning to face.

1. If machines can generate convincing language, how do we protect truth?
2. If they can influence opinion, how do we guard human judgment?
3. If they can imitate creativity, how do we define originality?

These are not technical questions alone. They are social, ethical, and deeply human. It is likely that the answers will evolve slowly, shaped by laws, culture, and the daily choices of ordinary people.

What feels clear already is that we have crossed a threshold.

Thinking with Machines

We once imagined a future where machines might learn to speak. Now, we are living in that future. The conversation has begun.

Perhaps one day, machines may not only speak with us, but work alongside us in ways that reshape science, medicine, education, and art. Or perhaps their greatest role will remain what it is today. A tool that extends our thinking without replacing it.

As I write this, I am aware of the strange loop we now inhabit. A machine helps us understand machines. A creation reflects its creator.

And as in every chapter of this book, the story circles back to the same source.

It all began with a thought.

Epiphanies from This Wave

PART 3 – Digital Tsunami

— *A stone tugged at iron, revealing an invisible pull. That secret grew into compasses, telegraphs, and at last voices riding unseen waves.*

— *Ada Lovelace saw not just numbers, but poetry in calculation. She dreamed machines could weave logic as a loom weaves silk.*

— *Sand, once scattered beneath our feet, was forged into silicon chips. In that alchemy, thought itself learned to live inside circuits.*

— *Navigation led to connection. Connection stitched the world together in wires and waves.*

— *"How can I help you?" whispered a machine. A human answered. The reply came back and, in that echo, machines felt alive.*

PART 4
Health, Humanity, and Science of Life

Chapter 1
Exploring the Human Body: A Journey Through the Centuries

When you step into a modern hospital, you enter a working map of knowledge. Monitors blink like small constellations, surgeons move with quiet focus, and data travels in digital streams faster than blood. Each specialist cardiologist, neurologist, orthopedist studies a corner of the same mystery: what the human body is and how it keeps itself alive. Behind this technical clarity lies a much older story, one of wonder, fear, and the first questions that pushed people to ask, what are we made of?

In the Beginning: Observing Shadows, Seeking Spirits

Long before white coats and laboratories, healing belonged to priests and elders. In ancient Egypt, illness was often read as an imbalance in the world or an offense against the gods. Still, beyond the ritual and prayer, careful records were kept. Texts such as the Ebers Papyrus cataloged symptoms, treatments, and surgical notes with systematic attention. People watched what happened, wrote it down, and learned.

Across the Fertile Crescent and the Indus Valley, clay tablets recorded remedies for fever and broken bones. In India, early systems of medicine developed frameworks that sought balance in the body's processes. In China, healers listened to the pulse and worked with the idea of Qi, an invisible flow that could be guided to restore health. These traditions did not dissect flesh as modern science would, but they observed, compared, and refined practices over generations. In time, observation quietly replaced pure superstition.

The Word Doctor: From Teaching to Healing

The title doctor comes from the Latin verb docere, to teach. The earliest physicians were scholars as much as they were healers. They gathered knowledge, tested remedies, and passed their lessons on. By the medieval period, a Doctor of Medicine was a figure who combined practical care with learned inquiry. This role mattered because medicine only advances when we ask why and then share what we find.

The Greeks Enter the Body: Thought Before Blades

The Greeks shifted the conversation. Hippocrates urged physicians to look at illness as a process rather than a divine punishment. His model of humor, blood, phlegm, yellow bile, and black bile was an attempt to describe balance in the body. Later, Galen dissected animals and extended anatomical theories that would dominate medicine for many centuries. His specifics were not always correct, but his insistence on testing and explaining bodies moved medicine away from pure superstition. Even an error can be valuable when it prompts further questioning.

The Renaissance: When Artists and Scientists Touched Flesh

For centuries, anatomy was constrained by custom and fear. Then the Renaissance reawakened curiosity. Leonardo da Vinci combined art and anatomy, sketching muscles and tendons with remarkable clarity. Andreas Vesalius went further; he dissected human bodies and published De Humani Corporis Fabrica, a work that reshaped anatomical knowledge. In the seventeenth century, William Harvey traced the circulation of blood, showing that the heart pumps in a continuous loop rather than allowing blood to wander aimlessly. The human body began to look less like a mystery and more like an organized system. Wonder remained, but it was now guided by evidence.

The Microscope Unveils a Hidden World

A new set of tools opened another dimension. Antonie van Leeuwenhoek, peering through lenses he had ground himself, saw life within drops of water: cells, tiny organisms, and structures invisible to the naked eye. The idea that the body contained microscopic actors, some of them harmful, changed everything. Louis Pasteur and Robert Koch later demonstrated that disease could come from micro-organisms, transforming medicine from ritual to prevention. Vaccines, antiseptics, and eventually antibiotics followed. Medicine learned to fight what it could now see.

Electric Nerves and Speaking Minds

The eighteenth and nineteenth centuries tested the idea that life was electrical and physical. Luigi Galvani showed that a spark could move a frog's leg, suggesting that nerves conduct electricity. Researchers began to map functions to structures. Paul Broca identified a region of the left frontal lobe tied to speech, giving language a place in the brain. These discoveries did not replace questions about consciousness, but they offered a new vocabulary: the mind has anatomy.

The Twentieth Century: Genes, Machines, and Light Without Scalpels

A decisive shift arrived in 1953, when James Watson and Francis Crick described DNA's double helix. The discovery revealed that life carries information in molecular code. Suddenly, biology had an instruction set to read and, eventually, to edit. At the same time, imaging technologies advanced. X-rays had already given clinicians a way to see bones; later, CT scans and MRI allowed specialists to look within soft tissue without cutting the skin. Surgery began to rely on precision guided by images. Diagnosis became a dialogue between machines and clinicians.

Where We Are Now: A Continuum, Not a Conclusion

The history of medicine is not a ladder that climbs toward a final summit. It is a river with currents that double back, converge, and branch. From herbal recipes written on clay to robotic arms that make precise incisions, our understanding of the body has shifted from the mystical to the measurable. Today, we combine genetics, imaging, data science, and clinical experience to treat patients more effectively than ever before.

Yet progress has not erased mystery. Genomics can map sequences, but it cannot, at a glance, explain why identical twins sometimes follow different health paths. Artificial intelligence can suggest diagnoses, but judgment still matters. Each heartbeat remains both a physiological event and a reminder of something larger. The body speaks, and we are still learning its language.

Listening as a Method

If there is a lesson in this long story, it is the importance of listening. Early healers listened to patients, watching the pulse, the color of the skin, the cadence of breath. Later, scientists listened to the evidence revealed by microscopes and imaging machines.

Today, listening includes data streams, genetic variants, and patterns learned by algorithms. Yet the human skill of attention, of noticing what changes, what does not, remains central.

A Personal Note

When I walk through a hospital corridor, I am struck by how recent some of these tools are. The fact that we can peer into a living brain or read a genome is astonishing, and yet it sits beside rituals as old as human civilization. I find that contrast humbling. The history of medicine teaches a kind of humility: new tools extend our reach, but they do not remove the responsibility to observe, to question, and to care.

The story continues. New therapies, edited genes, and smarter machines will reshape medicine further. But the heart of healing is unchanged. We listen, we learn, and we try again. That persistence, more than any single discovery, *is the true miracle of medicine.*

Chapter 2
Charles Darwin – The Man Who Let Nature Whisper

The greatest discoveries often arrive without fanfare. They come in moments of quiet attention, when someone looks longer than others, listens more closely, and refuses to dismiss the small detail. Charles Darwin belongs to that modest order of thinkers. He did not storm into knowledge on the back of dramatic experiments. He walked into it, notebook in hand, attentive to the world as it presented itself. He listened.

Charles Darwin
Photograph by Julia Margaret Cameron, c. 1868

Early Life and the Habit of Wonder

Darwin was born in 1809 in Shrewsbury, England, into a family where curiosity and learning were part of daily life. His grandfather, Erasmus Darwin, had already mused about change in living things, but young Charles was not raised as a prodigy. He preferred open fields to lecture halls, and beetles to textbooks. He had, above all, a patient delight in collecting and noticing. That habit, more than formal schooling, shaped his mind.

He tried medicine, then theology, and in each step the world beyond rooms and books kept tugging at him. At Cambridge, he met John Stevens Henslow, a botanist who saw in Darwin a rare attentiveness. Henslow recommended him for a voyage on the HMS

Beagle. The invitation was simple; the consequence was profound. Darwin set sail not to overturn received truths, but to record and to wonder.

The Voyage That Changed a Life

The Beagle's journey lasted nearly five years. From the beaches of South America to the volcanic outcrops of the Galápagos, Darwin collected specimens, sketched landscapes, and wrote pages of careful notes. What stands out is how often he stopped to compare, to measure, to hold two stones or two beaks side by side. He noticed, for example, that finches from different islands had beaks suited to slightly different diets. Those differences were not random quirks, he began to suspect; they spoke of a deeper connection between organism and environment.

The scenes he recorded were sensory and concrete. He wrote of the color of a beetle's shell, the angle of a cliff, the way certain plants clustered on windswept slopes. He described bones and shells with the practical eye of a collector and with the slow empathy of someone learning a language spoken by the living world. He returned to England not with a finished theory but with questions that would not let him go.

From Observation to Idea

For nearly twenty years, Darwin worked quietly. He organized his notes, re-examined specimens, and read voraciously. He tried to fit patterns into frameworks, and he tested explanations against evidence. That long, careful labor is important to remember: the scientific revolution rarely happens as a sudden flash. More often, it is the product of patient folding of fact into thought.

The central insight that emerged was simple and unsettling: variation exists within species, and some variations confer advantages in particular environments. Over many generations, those advantages can accumulate, forming new kinds. Natural selection, as a phrase later described it, is not cruelty or design; it is the process by which certain traits persist because they help organisms survive and reproduce in their circumstances. Fitness, in this sense, is not power; it is suitability to context.

It is also vital to see how Darwin framed his idea. He gathered evidence, fossils, geographic distributions, anatomical similarities, and let those facts speak. He did not leap to grand metaphysical claims. He offered a mechanism that could explain how complexity and diversity arise from simple processes repeated across deep time. That restraint, that devotion to evidence, shaped how he presented his work.

Publishing and the Moment of Reckoning

When Alfred Russel Wallace, working independently in Southeast Asia, arrived at a similar conclusion, Darwin felt the urgency of publication. In 1859, he published On the Origin of Species. The book did not thunder. It argued gently, methodically, drawing the reader through observations into an explanation.

That was a strategic choice. Darwin trusted readers to follow evidence rather than be shocked into belief.

Responses were mixed, as one might expect. Many scientists embraced his method and his data. Many religious thinkers felt alarmed. Others tried to find a way to reconcile observation with belief. Public debate was sharp, sometimes personal, and always consequential. Darwin did not seek to overthrow faith; he sought to reframe how we understand life. His writing, careful and modest, made the conversation possible.

A Life of Work and Quiet Courage

Darwin did not relish the spotlight. He withdrew to Down House, where his days were structured around observation, correspondence, and reflection. Illness visited him frequently, but it never deterred his attention. He studied orchids and barnacles with the same meticulous patience he had shown in the Galápagos. He explored questions about the expression of emotion across species, and he tested his ideas in smaller, concrete studies that strengthened the larger argument.

He also faced public trials. Thomas Huxley, often called Darwin's champion in debate, took on critics with fervor. Darwin himself preferred demonstration to rhetoric. He revised his views as new evidence arrived, and that intellectual humility became part of his legacy.

What Darwin Gave Us

The impact of Darwin's thought is immense and multi-dimensional. Scientifically, it provided a framework that allowed later discoveries of Mendelian genetics, DNA, and molecular biology to cohere into a deeper understanding of heredity and change. Philosophically, it shifted humanity's self-image. We ceased to be isolated artifacts and became participants in an evolving world. Ethically and culturally, the implications were vast and messy, provoking debate that continues to this day.

But perhaps the deepest gift was methodological. Darwin showed how patient observation, careful comparison, and slow reasoning can reveal processes that run on timescales far longer than a human life.

He taught us to notice the ordinary and to recognize patterns that, once seen, rearrange how we understand everything else.

A World Reframed

Before Darwin, many people assumed species were fixed, created in their present form. Fossils were curiosities, not clues. After Darwin, variation, adaptation, and deep time became central concepts. Life was understood as a dynamic process, a branching tree rather than a fixed chart. That change in perspective altered not only biology but other fields too. When we accept that forms change, we begin to imagine that ideas, cultures, institutions, and morals might also be subject to historical forces of transformation.

When I read Darwin, I am struck less by the man than by the method: look, compare, and keep looking until a pattern insists. That method is a kind of kindness toward the world. It refuses quick judgments and rewards patience. If there is a practical lesson for us, it is this: in our rush to conclusions, we often miss the small differences that matter. Listen, and you will find the world speaking in a language of slow nuance.

Darwin did not pronounce a final answer to the meaning of life. He offered a robust explanation for how life changes and diversifies. He left open questions and invited future inquiry. Today, as we manipulate genomes, simulate ecosystems, and reconstruct deep histories, we are walking in a path he helped clear. The boy who chased beetles taught us to read the living manuscript of nature. He showed that the ordinary contains hints of the extraordinary, and that quiet attention can change our view of everything.

Charles Darwin heard nature's whisper and translated it into a language we could all read. That act of translation remains one of the most profound gifts of science: it turned wonder into understanding, and in doing so, it reshaped how we see our place in the world.

Chapter 3
The Thought of Overcoming Limits

I walk along a quiet hospital corridor. Machines breathe in the half-light. Monitors blink like distant stars. Voices travel softly along the walls, measured and deliberate, careful not to disturb the fragile work of repair underway. In that stillness, I sense something familiar, something present in every era of healing: a single question, quietly but persistently pressing upon the human mind. When did we stop surrendering to mystery and begin reading the body as a text?

There was a time when illness was interpreted as an omen rather than a condition. Fever appeared as a divine message, pain as punishment, and infection as moral failing. We feared the body because we did not yet understand it. Then curiosity began its slow and disciplined work. One question at a time, humans learned to translate flesh into meaning. A shiver was no longer arbitrary; it had a cause. A wound followed a process of injury and repair. Fear gave way to fascination, and fascination matured into knowledge.

Transformation rarely arrives through spectacle. It emerges through patience, through observation sustained by minds unwilling to accept ignorance as final. In the seventeenth century, a physician named William Harvey watched the pulse until it ceased to appear miraculous and began to reveal its order. He traced the movement of blood as one would trace a river, discovering that it did not disappear into mystery but returned in a continuous circuit. The heart was no longer only a symbol of feeling. It became an engine of motion. With that shift, metaphor yielded to mechanism.

The microscope, developed and refined across generations, expanded perception further. Antonie van Leeuwenhoek, a tradesman driven by persistent curiosity, turned finely ground glass toward a drop of pond water and discovered a teeming world of microscopic life. He was not searching for meaning or salvation. He was simply looking. What he revealed transformed our understanding of disease. Fever and infection ceased to be purely spiritual judgments and became interactions among living organisms too small to see. At its best, science is not certainty but humility before the unseen.

Each lens we polish and each layer we uncover reveals not final answers but new questions. The pattern endures. In the twentieth century, James Watson and Francis Crick advanced this tradition not by refining instruments but by assembling ideas. Their model of the double helix rendered inheritance legible. DNA transformed fate into information. Disease became a sequence to be read and, perhaps one day, revised. Illness appeared less as destiny and more as code.

Then the tools evolved once again. Glass yielded to algorithms. Notebooks gave way to vast fields of data. Machines began to recognize patterns beyond the reach of human perception. Artificial intelligence extended the diagnostic gaze, scanning thousands of images in the time it takes a physician to blink and detecting subtle signals that precede harm. This is not a replacement. It is amplification. The stethoscope now listens with silicon ears.

Consider the small devices on our wrists. They trace heartbeats, measure sleep, and send a constant stream of data to clouds we cannot touch. Those signals can warn a patient and guide a clinician, but they also raise urgent questions about trust, ownership, and privacy. Who guards the record of your pulse? Who decides how that rhythm will be used? The same connection that saves a life can expose the most intimate traces of it. Connectivity demands a new kind of ethics, a new form of responsibility.

And then there are the surgical hands of machines. Robotic arms perform incisions with sub-millimeter accuracy, reducing trauma and recovery time. Gene editing offers precise edits to DNA that once would have seemed like miracles. CRISPR technology lets scientists rewrite tiny stretches of the genetic script, correcting errors that cause disease. These are real interventions that improve lives, but they also force us to ask deeper questions about what it means to alter the blueprint of human biology.

Progress has not been a straight line. It is a circulation, looping through failure and revelation, through loss and renewal, much like the blood it studies. For every triumph, there have been missteps and harms. Vaccines saved millions and were sometimes delayed by fear and politics. Surgical advances have reduced suffering for many and created new dilemmas for others.

Each step forward requires humility and care.

There is a pattern here that matters beyond medicine. Consider the shift from miracle to mechanism, from story to system, from awe to method. It is the same movement we observe in the technologies that move us, the narratives that shape us, and the habits that govern us. The mind that insists on explanation is the mind that can act. To treat is to intervene in a system, and to intervene responsibly is to know more and to ask better questions.

Perhaps the most important lesson is not that we will ever conquer death, but that we continue to refuse finality. We keep asking even when answers seem small or slow. The story of medicine is not dominion over nature. It is a conversation with it: a continual negotiation, an exchange where curiosity meets compassion.

Walking back down the corridor, I hear the soft hum of machines again. In that sound, I hear continuity: Harvey watching pulses, van Leeuwenhoek peering at water, Watson and Crick sketching spirals, engineers building tools that extend human touch. We are heirs to a long line of simple refusals to accept the limits of knowledge. That refusal is not arrogance; it is persistent attention, the slow art of listening closely enough to learn.

We cannot predict every outcome. We should not promise certainty where none exists. But we can keep refining our instruments, our ethics, and our questions. The body keeps speaking. We keep listening. And in that ongoing exchange, humanity continues to heal itself.

If there is a miracle in this story, it is modest: not the triumph over all illness, but our refusal to stop asking what lies within.

Chapter 4
Invention of Vaccine

The Thought of Disease Prevention: The Birth of Vaccines, Antibiotics, and Hand Washing

Every great turning point in human health began not with advanced machines or towering laboratories, but with a simple and courageous question: Could this suffering have been prevented? That quiet question reshaped how we understand life, illness, and survival. From it emerged one of humanity's most powerful ideas, that invisible forces shape our fate, and that even the smallest human actions can tip the balance between life and death.

For most of history, disease was a mystery wrapped in fear. People blamed curses, bad air, or divine punishment. Death arrived without warning, and no one truly understood why. Yet, across centuries, a few minds dared to suggest something radical. Perhaps illness followed rules. Perhaps it could be studied. Perhaps it could even be stopped.

That shift in thinking did not arrive all at once. It unfolded through quiet observations, stubborn resistance, accidental discoveries, and an unwavering belief that human beings could learn to protect themselves.

The Unseen Killer: The Doctor Who Saw the Invisible

In the mid-nineteenth century, hospitals were meant to be places of healing. Yet, inside many of them, death walked freely. At the General Hospital in Vienna, a young Hungarian physician named Ignaz Semmelweis noticed something horrifying. Women were dying in childbirth at alarming rates. Even more troubling, the death rate was far higher in wards attended by doctors than in those run by midwives.

The difference made no sense at first. Both wards treated the same condition. Both served the same community. Yet one was far more deadly than the other.

Semmelweis watched carefully. He paid attention not to the grand procedures, but to the smallest habits. Doctors moved from dissecting corpses in the autopsy room to delivering babies, often without washing their hands. Midwives did not perform dissections. They moved directly to care.

Semmelweis asked a question few were willing to ask. *What if* something invisible was being carried on the hands of the doctors themselves?

At the time, germs were unknown. There was no microscope powerful enough to reveal bacteria. Yet Semmelweis trusted what the numbers were showing him. He ordered doctors to wash their hands in a chlorinated lime solution before touching patients. Within weeks, maternal deaths fell dramatically.

The result should have been celebrated. Instead, it was rejected.

To accept his discovery meant accepting that doctors themselves had been the cause of countless deaths. Many could not bear that truth. Semmelweis was mocked, resisted, and isolated. Eventually, he was committed to an asylum, where he died from an infected wound, killed by the very unseen enemy he had tried to expose.

Years later, Louis Pasteur would confirm what Semmelweis had sensed through evidence. Germs existed. They caused infection. Joseph Lister would apply this knowledge to surgery, using antiseptics to drastically reduce deaths in operating rooms.

Today, every scrubbed hand before surgery, every sanitizer dispenser in a hospital corridor, carries the legacy of Semmelweis. His insight changed medicine forever. Clean hands became a shield.

The Birth of Vaccines: Training the Body to Remember

Long before the science of microbes existed, people noticed something curious. Those who survived certain diseases often did not catch them again. In parts of China and India, healers practiced variolation, deliberately exposing people to mild forms of smallpox in hopes of building protection. It was risky, but it hinted at a deeper truth. The body could learn from illness.

In 1796, a quiet country doctor in England named Edward Jenner made a leap that would change the world. He observed that milkmaids who had once caught cowpox rarely developed smallpox. Cowpox caused only mild illness. Smallpox was often lethal.

Jenner asked a daring question. Could exposure to cowpox protect against smallpox?

To test this, he introduced material from a cowpox sore into the arm of a young boy named James Phipps. Later, he exposed the boy to smallpox. The child did not fall ill. Against a tradition of fear and superstition, Jenner had shown that the body could be trained to defend itself.

The first vaccine was born. The word vaccine comes from vacca, the Latin word for cow. From one small observation in a rural village came a tool that would one day eradicate smallpox entirely and push diseases like polio close to extinction.

Louis Pasteur later expanded this principle, developing vaccines for rabies and anthrax. He proved that microorganisms caused disease, not curses or bad air. With each new vaccine, humanity gained not just protection but confidence. We were no longer entirely at the mercy of nature. We could prepare for it.

Vaccination transformed childhood from a battlefield into a safer beginning. It extended life expectancy, stabilized populations, and redefined what prevention could mean.

The Accidental Revolution: The Birth of Antibiotics

Some of the most powerful ideas in medicine did not arrive through careful planning, but through alert curiosity. In 1928, Scottish bacteriologist Alexander Fleming returned from a short holiday to find one of his laboratory culture plates contaminated by mold. Around the mold, the bacteria had vanished.

Many researchers would have thrown the sample away. Fleming paused. He asked why. He realized the mold was producing a substance that killed bacteria. He named it penicillin.

At first, penicillin was difficult to isolate and produce. For years, it remained a laboratory curiosity. Then, during the Second World War, the urgency of battlefield injuries pushed research forward. Scientists Howard Florey and Ernst Chain found ways to purify and mass-produce the drug.

Suddenly, infections that once guaranteed death became treatable. Soldiers survived wounds that would have killed them in earlier wars. After the war, antibiotics spread rapidly into civilian medicine. Hospitals transformed. Recovery replaced mourning. Surgery became safer. Childbirth became less perilous.

More antibiotics followed. Streptomycin, tetracycline, and many others arrived, each saving millions of lives. For the first time in history, humanity could fight bacteria with precision.

Yet this victory carried a warning. Bacteria evolve. Overuse of antibiotics allows resistant strains to develop. Even today, researchers pursue new solutions with renewed humility, aware that the struggle between microbes and medicine is not a war with a final ending.

A Thought That Guarded Civilization

From disinfected hands to vaccinated bodies and antibiotic treatments, disease prevention reshaped the human story. It extended lifespans, stabilized societies, and allowed civilizations to grow without being constantly torn apart by epidemics.

Each breakthrough began with observation. Each faced resistance. Each demanded the courage to challenge old beliefs.

Hand washing reduced invisible danger.

Vaccines taught the immune system to recognize its enemies.

Antibiotics turned fatal infections into survivable conditions.

Together, they redefined the relationship between thought and survival. Knowledge became protection. Curiosity became a form of defense.

As I reflect on these stories, one truth stands out with clarity. Progress in medicine has rarely come from certainty. It has come from doubt. From the willingness to look at suffering and ask whether it must exist at all.

Every clean hospital ward, every childhood immunization, every life saved by timely antibiotics is a continuation of those early questions. We are the beneficiaries of minds that refused to accept death as inevitable.

The fight against disease is not over. New viruses emerge. Resistance grows. Global health remains fragile. Yet the central lesson endures. When human thought is guided by compassion and discipline, it becomes one of the most powerful medicines the world has ever known.

In the quiet routine of washing hands.

In the careful science of vaccines.

In the unlikely mold on a forgotten dish.

Thought did not merely observe life. It learned how to defend itself.

Chapter 5
Pain-Free Surgery: The Birth of Anaesthesia

From Agony to Question

Not long ago, the operating room was a chamber of endurance. Patients were strapped to tables and expected to bear the knife as a test of courage. Surgeons moved with speed because speed was mercy; the quicker the cut, the less time for suffering to register. The air in those rooms held fear and iron, groans and the scent of unwashed linen. I imagine a man grasping for breath while voices barked orders. I imagine a family waiting in the corridor, counting time like a kind of cruel prayer.

Then someone asked a simple but daring question: could pain be silenced? Could the body be opened and the mind remain whole? That question changed medicine more than any single instrument ever did.

Early Attempts and Folk Remedies

Long before modern science, healers tried what they could. Across distant civilizations, people mixed bitter brews, chewed herbs, and wrapped the injured in rituals of comfort. Opium eased pain in ancient Egypt. Plant extracts and fermented drinks were used in China and India to dull sensation. These remedies offered moments of relief, not solutions. They were fragments of mercy scattered across geography and time, each one a small step toward a larger idea.

A Laughing Revelation

The real turning point came in a place you would not expect: a stage where a crowd laughed at a showman. In the 1840s, a dentist named Horace Wells sat in the audience while someone inhaled nitrous oxide, laughing gas. The man fell and struck his leg, but cried out no more. Wells saw a possibility where others saw a joke. He tested the gas on himself, had a tooth removed, and felt nothing but surprise and relief. He believed he had found a way to remove suffering.

Wells tried to demonstrate his discovery to his professional peers in Boston, but the trial failed. The patient cried out, and the audience jeered. Humiliation followed.

Wells retreated, his idea dismissed and his reputation damaged. Yet the thought itself did not die. A seed has been planted, and ideas have a way of growing where they are needed.

Ether Day and the First True Demonstration

A more decisive moment arrived when another dentist, William Thomas Green Morton, pursued the promise with careful experiments. Morton found ether to be a powerful anesthetic. On October 16, 1846, at Massachusetts General Hospital, he placed a glass inhaler to a patient's lips. Surgeon John Warren began the procedure. The room waited. There was no cry, no flinch. When the operation was finished, Warren turned to the crowd and declared, Gentlemen, this is no humbug. The silence in that moment was not empty; it was profound. The age of agonized surgery had ended. Physicians later called that day Ether Day.

Chloroform, Controversy, and Compassion

Progress did not follow a straight line. In Edinburgh, James Young Simpson experimented with different vapors and found chloroform to be a reliable and gentler agent than ether for many uses. Yet medical advances often run headlong into social belief. Some religious leaders argued that pain, especially in childbirth, was ordained or purifying. Simpson answered not with theology but with practice. He insisted that relieving suffering was an act of kindness. His conviction found a public ally when Queen Victoria accepted chloroform during the birth of her child in 1853. When the monarch endorsed relief, public resistance faded. A private practice in a parlor became public medicine.

From Novocaine to the Modern Theatre

The story did not stop at ether and chloroform. By the early twentieth century, chemists had synthesized procaine, known commercially as Novocaine, which brought precision to dentistry and minor operations. Later developments introduced intravenous agents, refined inhalation mixtures, and monitoring devices that monitor oxygen, heart rhythm, and breathing. Anaesthesiology grew into a specialty, a discipline of balance. Today, an anaesthesiologist acts like a conductor, keeping the patient in a careful state: asleep enough to feel no pain, awake enough to preserve life's essential functions.

Technology and the Quiet Art of Control

Modern anaesthesia is technical and humane. We now have machines that deliver precise drug doses, monitors that alert us to minute changes, and techniques that target nerves so operations can be done with local rather than full sedation. We can control pain, regulate blood pressure, and ensure that the body's systems remain steady while surgeons work. The operating room, once a place of desperate endurance, has become a controlled environment where fear is minimized and outcomes are maximized.

Ethics and the Measure of Progress

The move from suffering to relief is not only a story of chemistry and skill. It is a story of values. To ask whether we should remove pain is also to ask what we owe to one another. When practicing anaesthesia became possible, it forced society to confront a moral choice. Is it compassionate to spare another person pain if we can? For those who believed in the duty to alleviate suffering, anaesthesia was a triumph of morality; for others, it raised anxieties about unnatural intervention. The debate recedes now into medical practice, but the underlying question remains: progress is always linked to how we choose to use it.

A Quiet Revolution of Empathy

What began as a desperate wish must we suffer to be saved became a field of science and a measure of human compassion. The path from crude potions and accidental discoveries to the precise, watchful craft of modern anaesthesiology shows how empathy and curiosity combine to reshape our world. Each silent operating room holds the echo of those early questions and the gentle answer they produced.

I return to these stories because they remind me, and I hope they remind you, that some of the greatest advances begin as small rebellions against what we accept as inevitable. Pain was once seen as part of life's price. Today, in a theatre where a child sleeps without fear and a mother gives birth with comfort, we see the result of a question asked boldly enough to change practice and policy. That question still matters: how will we use our knowledge to make life more humane?

The birth of anaesthesia is proof that medicine can be both exacting and kind. It teaches us that progress is not only technical mastery; it is the steady refusal to accept suffering as normal when relief is possible.

Chapter 6
Seeing the Invisible: The X-Ray and MRI Revolution

There was a time when looking inside the human body meant a surgeon's blade and the patient's courage. Doctors probed with hands and tools, guided by experience and luck, hoping their touch would find the problem. The idea of seeing beneath skin without opening it sounded like a miracle or a magician's trick. Yet curiosity keeps knocking at closed doors, and someone somewhere refuses to accept that the unseen must remain secret.

The glowing hand

In late 1895, Wilhelm Conrad Roentgen was working alone in a dim laboratory, experimenting with cathode rays and photographic plates. One night, a strange glow appeared on a plate. When he developed it, he saw the outline of a hand, bones black against pale tissue, and a bright ring marking a wedding band. Roentgen called the rays X, for unknown reasons. He had found a way to make the invisible visible.

The discovery moved with astonishing speed. Within months, surgeons used X-rays to locate fractures and foreign objects. For the first time, diagnosis could happen without an incision. Roentgen declined to patent his finding; he shared it. In 1901, he received the first Nobel Prize in Physics. Medicine's vision had changed forever.

Beyond the bones

X-rays reveal structure. They show bone, metal, and dense tissue, but they do not reveal the soft subtleties of living organs, the motion of blood, or the shifting chemistry of disease. To see deeper requires listening to the quieter songs of matter itself.

In the mid twentieth century, physicists discovered that atomic nuclei respond to magnetic fields with tiny signals. This phenomenon, nuclear magnetic resonance, was described by Felix Bloch and Edward Purcell. Their work, which earned them a Nobel Prize in 1952, became the scientific foundation for later imaging techniques.

Over decades, engineers and physicians asked a new question: could these faint signals be translated into pictures of living tissue?

Several researchers contributed pieces of the answer. Raymond Damadian observed that healthy and diseased tissues produced different magnetic responses and proposed that this difference might reveal tumors. Paul Lauterbur showed how applying magnetic field gradients would allow those responses to be mapped into images. Peter Mansfield developed mathematical methods that made image acquisition faster and more practical. Their combined work converted an abstract signal into a readable map of the body.

By the late 1970s, magnetic resonance imaging became a reality. Early scanners produced images with a strange and haunting clarity, revealing soft tissue, the gray folds of the brain, and the subtle contours of organs without cutting or ionizing radiation. MRI offered not only anatomy but also function. With refinements such as functional MRI and diffusion imaging, physicians could watch blood rush through a tumor, trace the pathways of water in neural tissue, and see how the brain lights up with thought.

From vision to precision

The arc from Roentgen's glowing hand to the living maps of MRI is a story of questions that would not be satisfied. Computed tomography, developed shortly after midcentury by Allan Cormack and Godfrey Hounsfield, added three-dimensional clarity by combining X-ray slices into volumes. Positron emission tomography allowed physicians to trace metabolic processes by following labeled molecules. Ultrasound turned sound echoes into moving images in real time. Each technology brought a new way to see, and each built on the restless refusal to accept limits.

These inventions changed medicine's posture. Diagnosis moved from guess and cut to observation and precision. Treatments grew more targeted because surgeons and physicians could plan with sight. Diseases that once required exploratory operations could be found with a scan in minutes.

A personal note

When I first studied these developments, I felt something like awe and gratitude. Here, human curiosity turned into mercy. A physicist's late-night surprise and the slow, patient work of engineers gave millions a chance to avoid unnecessary surgery and receive earlier, better care. It reminds me that science is not merely an accumulation of facts. It is also a chain of human moments: a question, a stubborn experiment, an argument in a corridor, a failed attempt, and then a small success that changes many lives.

Ethics and limits

Seeing more also asks us to think more. Scans expose intimate details that were once private. They reveal incidental findings that can worry patients and spark difficult decisions. The power to look demands responsibility in how we use information and share results. Technology brought vision, but wisdom must guide the gaze.

In the end, the X-ray and MRI revolution did more than open bodies. It altered the relationship between curiosity and care. It showed that the boundary between what we cannot see and what we can know is not fixed. When a single question refuses to be ignored, the world yields a little of its secret. What began as a greenish glow on a plate became a new language of health, one that speaks without a blade and listens without intrusion.

Chapter 7
The Rise of Allopathic Medicine

When we fall ill today, most of us reach for a tablet, call a doctor, or walk into a hospital whose bright corridors promise relief. That quiet confidence in modern medicine did not appear overnight. It is the harvest of centuries of questions asked in candlelit rooms, of failures turned into breakthroughs, and of minds unwilling to accept suffering as inevitable. The story of allopathic medicine, the branch that came to define modern Western practice, is the story of how curiosity learned to heal.

Many Streams of Healing

Long before the first stethoscope touched a chest, healing was a practice woven from observation, ritual, and close listening to the natural world. In India, Ayurveda taught the importance of balance among bodily systems and the role of diet and lifestyle. In China, practitioners developed systems centered on the flow of Qi and the harmony of complementary forces. In the Mediterranean and Middle Eastern worlds, the ideas of Hippocrates evolved into Unani medicine and later found refinement in thinkers such as Avicenna, who combined empirical observation with philosophical reasoning.

Other regions preserved their own traditions of plant knowledge, careful observation, and communal care. For all cultures, the aim was the same: to lessen pain and restore what illness had disturbed. These traditions were not accidental. They were long experiments in empathy, the accumulation of generations of careful watching and testing, recorded in the language of plants, seasons, and human experience.

The Allopathic Turn: From Intuition to Investigation

Then, as Europe moved through the Renaissance, attitudes toward disease began to change. People stopped accepting the unseen as the final truth and began to demand evidence. Illness stopped being explained only as divine punishment and became instead a subject for inquiry. Out of that shift emerged the core idea behind allopathy: disease might be understood and treated by applying its opposite, by manipulating the body or environment in ways that produced measurable improvement. This approach did not immediately discard older traditions, but it insisted that belief be checked against observation.

A Chain of Groundbreaking Discoveries

From this insistence on evidence came a cascade of revelations. Each discovery was built on careful observation and patient experiment.

William Harvey mapped the circulation of blood, revealing the heart as a pump propelling a closed system of vessels. That insight reframed how physicians imagined the body's internal economy.

Edward Jenner observed that milkmaids who contracted cowpox rarely developed smallpox. From that observation, he devised the practice of vaccination, a method that would later prove central to public health.

Louis Pasteur exposed the microbial world, arguing that tiny organisms could cause disease. His experiments overturned long-held assumptions and laid the foundation for germ theory.

Joseph Lister introduced antiseptic techniques into surgery, applying Pasteur's ideas to the operating room and transforming surgery from a desperate gamble into a controlled intervention.

Paul Ehrlich pursued the concept of a magic bullet, a compound targeted to attack pathogens without harming the host. His work foreshadowed modern pharmacology.

And by a combination of serendipity and attentiveness, Alexander Fleming noticed that a mold had inhibited bacterial growth, a discovery that led to penicillin and launched the antibiotic era.

Each of these moments was more than an isolated insight. They were instances of the scientific pattern: careful observation followed by hypothesis, experiment, failure, and then refinement. Thought, tested against the world, produced tools that changed millions of lives.

The Modern Era of Medicine

As instruments sharpened, so did our sight. Microscopes revealed cells. X-rays let us peer inside the living body. Magnetic resonance imaging and computed tomography produced maps of soft tissue and bone. The human body, once a mystery, became a landscape to be explored and measured.

Hospitals evolved from places of shelter and comfort into centers of precision. Laboratories and clinical trials created standards for testing treatments. Medicine moved from anecdote to evidence. Today, we edit genes, engineer replacement tissues, and use artificial intelligence to read patterns no human could see. Yet behind every high technology still burns the same spark that lit the earliest healer's fire: the simple, urgent question of how to make someone well.

Respecting Other Paths

Allopathic medicine commands much of the modern stage, but other systems persist and adapt. Many practitioners now draw from multiple traditions, combining tested methods from modern science with insights from long-standing practices where appropriate. Patients and clinicians often seek a balance between the rigor of clinical trials and the holistic attention found in older practices. Healing has never belonged to a single school; it is a chorus of human attempts to understand the body's needs.

The Thought That Heals

At its heart, medicine, ancient or modern, begins with a single question whispered across the ages: How can I help this person feel better? That question is the first remedy. It is the thought that heals.

We must remember that medicine advances not only through grand discoveries but also through modest, daily acts of care. Empathy, careful listening, and diligent follow-up are as much a part of healing as any technology. The scientific leaps give us tools, but the human impulse to relieve suffering gives them purpose.

Chapter 8
Blueprints of Ourselves: How Humans Recreated Their Own Bodies Through Machines

Human curiosity eventually turned inward. After centuries of borrowing patterns from the sky, the seas, and the animal world, we began to study the most intricate system we know, our own bodies. We did not simply seek to heal; we sought to replicate and extend. In that turning, our anatomy became a set of blueprints. Eyes inspired cameras, skin birthed sensors, hearts shaped pumps and circuits, and thinking inspired algorithms. This chapter traces that reflection, organ by organ, noting how machines echo our biology and, in doing so, reveal something deeper about what it means to be human.

Eyes to Cameras, Seeing Again

The eye taught us how to frame the world. Its curved lens, adjustable aperture, and layered retina suggested a design we could copy. Early lenses, crude and brittle, slowly became precise glass and then engineered optics. Modern sensors mimic rods and cones to extend our vision into the dark and across the electromagnetic spectrum. Autofocus reacts in microseconds like a pupil contracting, and image stabilization calms the tremor of a hand. Where once a human blinked to change focus, satellites and microscopes now grant sight to realms no eye could reach. The machine did not replace seeing; it multiplied it.

Ears to Microphones and Hearing Aids

The ear converts pressure into meaning, and the microphone turns that conversion into technology. Early microphones were clumsy and tinny, but the goal was always more than volume. Engineers wanted clarity and nuance, a way to preserve the whisper as well as the shout. Cochlear implants have done something profound. They do not recreate the ear perfectly, yet they restore the cadence of speech and the intimacy of a voice. In rooms where silence once ruled, human connection returns.

Nose to Electronic Sensors

Scent is a map of the body and the world. Humans long relied on smell to warn of danger and to find food. Today, arrays of chemical sensors, sometimes called electronic noses, can detect volatile compounds that indicate disease, spoilage, or contamination. They do not smell as we do, but they translate molecules into data. In clinical settings, these systems show promise in screening for certain cancers and infections from breath samples, a reminder that invisible information can be practical and urgent.

Voice to Speech Synthesis

When machines learned to speak, we did not merely program phonemes. We taught them to carry nuance. Early speech synthesizers were mechanical and robotic. Modern systems learn prosody, rhythm, and pause. They attempt not to imitate a voice perfectly, but to convey warmth, irony, or comfort. When a synthesized voice times a pause as a human would, the effect is not mere convenience; it is empathy expressed in code.

Skin to Electronic Touch

Touch anchors us to the world. Recreating it for machines required thin, flexible sensors that respond to pressure, heat, and texture. Electronic skin panels can now register the difference between a fingertip pressing and a hard strike. On prosthetic hands, these sensors provide feedback that helps a user adjust grip strength without crushing a fragile object. In robotics, that sensitivity makes machines safer collaborators. The goal is not to reproduce human feeling perfectly; it is to make machines feel enough to act gently.

Heart and Lungs, the Rhythm of Life Recreated

Mechanical pumps and oxygenators are perhaps the most literal translations of biology into engineering. Ventricular assist devices support a failing heart, and total artificial hearts sustain life while a transplant is possible. Ventilators and extracorporeal oxygenation carry breath across machines. These interventions do not merely replace function; they buy time and restore presence, letting families speak at bedsides that would otherwise have fallen silent. They are devices of survival and of profound human consequence.

Brain to Computers and Artificial Intelligence

The brain is a vast, poorly charted ocean, yet even fragments of its operation have inspired computational models. Neural networks mimic aspects of neural connectivity to recognize images, translate languages, and compose music. Brain–computer interfaces translate electrical patterns into cursor movements, letting a paralyzed hand control a screen with thought alone. These advances are fragments, not full copies. We should say perhaps or likely when we speculate about replicating consciousness. Still, each algorithm is a mirror that helps us inspect our own thought processes, and each interface affords dignity to people once locked out of interaction.

Hands and Legs, Motion Reborn

The hand converts intention into effect. Modern prosthetics translate residual muscle signals into fine motor control. Bionic hands grasp, pinch, and even perform delicate tasks under precise control. Powered exoskeletons return mobility to people who have lost it. The technology is not just engineering; it restores agency, the sense that one can act in the world again.

Bones and Muscles, Strength from Matter

When structure fails, engineers have learned to substitute materials that are lighter and stronger. Titanium and medical ceramics anchor implants that carry load and withstand decades of use. Artificial muscles based on smart polymers contract with electric stimulus, promising soft robotics that emulate natural movement with less noise and more adaptability than rigid motors.

Organs Engineered Within

Dialysis replaces kidney filtration, oxygenators take over lung exchange during complex surgery, and bioengineered tissues point toward organs grown or printed rather than harvested. These are not distant fantasies. They are incremental progress, each step subject to ethical questions and practical limits. Artificial wombs remain experimental, but they force us to reconsider where and how life begins and how medical technology intersects with society.

Blood, Vessels, and the Circulation of Invention

Synthetic grafts, microfluidic channels, and vascular scaffolds echo cardiovascular architecture at tiny scales. Engineers have learned that circulation is not only material flow but also design logic. A functioning blood substitute, or a reliable vascular scaffold for printed tissues, would reshape trauma care and transplantation. Here, biology and engineering meet at the level of flow and pattern.

The Mirror We Built

Across these technologies, one truth becomes clear. Machines reflect human priorities. They are not neutral instruments. In crafting cameras, sensors, motors, and algorithms, we reveal what we value: sight that can probe distance or magnify the tiny, hearing that restores conversation, touch that preserves dignity, and breathing that buys time. Technology is a conversation with nature. It is not a conquest. It is a translation.

When we look at these devices closely, we are looking back at ourselves. The machines extend our senses and our hopes. They reveal fragility and ambition, limitations and compassion. They demand questions of fairness, of access, and of purpose. Who receives these inventions? Who decides their design? What ethical limits should guide the desire to engineer life?

These are not rhetorical diversions. They are practical concerns. Development happens unevenly. Some technologies reach wealthy markets first, while basic needs remain unmet elsewhere. As we build more powerful tools, we must also build institutions and norms that ensure they serve broad human flourishing.

A Final Reflection

We have not escaped our biology by inventing replacements for it. We have learned to converse with it. The machines we create are maps of our wonder, and the choices embedded in their design tell the story of who we are. In the coming decades, we will press further into more sensitive prosthetics, deeper interfaces between brain and machine, and more lifelike tissue engineering. Perhaps some inventions will surprise us, and perhaps some will stall under ethical scrutiny. What is certain is that this conversation between flesh and machine will continue, and with it our obligation to ask not just what we can build, but why we should build it.

Chapter 9
Computers and AI in the Medical Field – Thought Meets Healing

For much of human history, healing was guided by instinct, experience, and the reassuring weight of a healer's hand. A pulse pressed beneath fingers, a careful glance, a voice that settled fear — these were the foundations of medicine. That intimate human rhythm still matters today. But beside it, a new kind of intelligence now works quietly. A mind built from silicon. A partner that does not sleep, does not forget, and sees patterns invisible to the human eye.

Computers and artificial intelligence have stepped beyond hospital machinery. They are now collaborators in care. The healer brings empathy, comfort, and wisdom. The machine brings speed, memory, and precision. Together, they create a field where diagnosis sharpens, surgery becomes more exact, and medicine learns to predict instead of simply react.

This partnership does not remove humanity from healing. It expands it.

Revolutionizing Diagnosis and Treatment

Understanding the illness is the first step toward recovery. In this area, artificial intelligence has become a powerful lens.

Algorithms trained on vast collections of medical images examine X-rays, CT scans, and MRIs with remarkable consistency. They detect tiny shadows of disease that even highly experienced clinicians sometimes miss. They do not grow tired at the end of a shift, and they never rush.

Digital systems read patterns in patient histories, treatments, and symptoms. They compare a single case to thousands of others, offering doctors a broader perspective. These tools do not replace clinical judgment. They strengthen it. Doctors still decide, but now with deeper insight than one mind could hold alone.

Redefining Surgery with Robotic Assistance

Step into a modern operating room, and you will see more than the hands of a surgeon. Mechanical arms may extend from a console, responding to the surgeon's smallest movements. Robotic assistance allows for smaller incisions, steadier techniques, and faster patient recovery.

Before the first incision, the software creates a virtual map of the patient's anatomy. Surgeons rehearse the procedure digitally, anticipating challenges before they appear. Human intuition guides the tools, yet machines support the precision. The patient benefits from both calm judgment and motion without tremor.

Patient Care Management: A New Kind of Listening

Healing does not end when surgery does. It continues through careful monitoring. Electronic health records now track information that once depended on memory alone. AI scans these records for early warnings, a slight shift in heart rhythm, a dangerous combination of medications, or a symptom pattern hinting at infection.

Some systems identify risk days before the body reveals obvious signs. Quietly, they remind nurses and doctors when action may be needed. This allows medicine to move from reacting to anticipating, giving patients a safer journey through recovery.

Virtual Healthcare Assistants: Helpful Companions at Home

Care now extends beyond the walls of a hospital. On a phone or computer, a virtual assistant can answer health questions, provide reminders, or help a patient explain symptoms. These tools reduce anxiety for people managing long-term conditions and offer comfort when the clinic is closed.

They do not replace nurses or physicians. They simply help patients stay connected to care. They are companions that check in at midnight or early dawn, when worry usually feels heaviest.

The Future of AI in Medicine: Where Thought Keeps Evolving

Despite the progress, we are still at the beginning. Algorithms will continue to improve, and new applications will emerge. Yet one principle remains clear: technology supports the healer, not the other way around.

Artificial intelligence can offer millions of comparisons in seconds. But it does not understand what a diagnosis means to a frightened patient. It does not feel the weight of a family's hope. That emotional understanding remains human territory.

The most advanced tool will never replace a gentle hand or a reassuring voice. It simply adds strength behind them.

A Shared Rhythm of Data and Care

Medicine is becoming a partnership between intuition and computation.

Between heart and hardware.

Between the wisdom of experience and the power of information.

We stand in a new era where data listens as closely as the healer does. Where machines enhance our ability to see illness early and treat it wisely. The future of care is not cold. It is more informed, more careful, and more human.

The healer still leads the way.

Now, technology helps light that path.

Chapter 10
Printing the Unprintable – Human Parts from a Machine

What if a machine could print a beating heart, not on paper or as a simulation, but as a living organ, each cell placed where it belongs, pulsing with human warmth? Once, that idea would have sounded like a myth. Today, it is science edging into reality.

I rest my hand on my wrist and feel that quiet pulse beneath the skin, steady and sure. That cadence, ancient and intimate, may one day be echoed by something built, not born. A heart shaped not by millennia of evolution but by human imagination. The question has shifted: not if this is possible, but how far we are willing to go.

This is not merely speculation. Bioprinting, the convergence of biology and engineering, is rewriting the boundaries of life. Like conventional 3D printing, a bioprinter builds layer by layer. Its "ink" is living material, a mix of cells, growth factors, and hydrogels that approximate the mechanical properties of tissue. The end product is not a toy. It is an attempt to recreate the structures that sustain life.

The process starts with imaging. A high-resolution CT or MRI scan captures the precise geometry of an organ, a valve, a piece of bone, or a section of liver. That scan becomes a three-dimensional map, the blueprint for the printer. The bio-ink is prepared, not from plastic or resin but from a soft gel seeded with stem cells. Sometimes those cells come from the patient, which reduces the risk of immune rejection. With microscopic precision, the printer deposits droplets of bio-ink, layer upon layer, sculpting tissue architecture and designing tiny vascular channels, the future highways for blood.

When the mechanical process stops, the biological process only begins. The fragile construct transfers to a bioreactor, a warm and controlled chamber that simulates the body's conditions. There it matures and reorganizes. Cells interact, the extracellular matrix forms, and, with luck, tissue gains function. In 2019, researchers in Tel Aviv printed a small, vascularized heart the size of a rabbit's. It was crude, but it beat. It was the first whisper of possibility. Elsewhere, scientists have produced skin tissues for burn victims, cartilage, tracheal segments, bone scaffolds, and kidney organoids that can perform limited filtration. These are not yet full, transplantable organs, but they are meaningful advances along a long path.

Imagine a world without long donor waiting lists. Imagine a child receiving a heart developed from her own cells. Imagine doctors testing drugs on a patient-specific liver printed at bench scale before exposing the patient to risk. Medicine would shift from one-size-fits-all to intensely personalized care, and clinical trials could become safer because treatments could be trialed on accurate models of a patient's tissue.

Yet enormous technical challenges remain. Chief among them is vascularization. Every cell in an organ needs oxygen and nutrients. Building a dense, functional vascular network that integrates with a patient's circulatory system is fiendishly difficult. Scientists can design microscopic channels and coax cells to line them, but reproducing the hierarchical complexity of arteries, arterioles, capillaries, venules, and veins is an engineering problem still in progress. Mechanical strength, innervation, and long-term viability are further hurdles.

Beyond technical questions, ethical and social questions loom large. Who owns a printed organ? Could life itself be patentable? Will these technologies widen medical access or deepen inequality? If replacement organs remain costly, the technology may benefit only the wealthy. If it becomes widely available, whole populations could see dramatic improvements in health and longevity. Those outcomes depend not only on science but on policy, on regulation, and on deliberate public choices.

We must also be candid about limits. Some milestones are likely in the near term, such as improved tissue grafts for reconstructive surgery and better organoids for research. Complete human hearts suitable for transplant in adults are a more distant prospect, likely requiring breakthroughs in vascular network engineering and long-term integration. I frame these developments gently, because the history of medicine is full of careful steps that accumulate into major change.

The implications extend beyond transplantation. Bioprinting reshapes drug development, toxicology testing, and the study of congenital defects. It opens the possibility of printing tissues with embedded sensors for monitoring, or scaffolds that release growth factors in timed sequences. It invites us to rethink the relationship between patient and treatment, between living tissue and engineered support.

If humanity has a pattern, it is this: first we express thought on paper, then we make it tangible, and finally we embody it. We printed words for centuries to spread ideas. Now we are learning to print parts of life itself. That transition is both wondrous and humbling. Every major technological advance carries responsibility. As we layer cells into tissues, we must also layer ethics, oversight, and equitable access into the field.

Every innovation begins as a thought. We once thought the printed page could transform knowledge. That thought did. We now harbor a new thought: that tissue printers might transform medicine. It will be a long journey from laboratory demonstrations to routine clinical practice. Along the way, we will celebrate achievements and confront hard moral questions. We will choose regulations, reimbursement models, and public priorities. We will decide whether this gift of creation belongs to everyone.

In the end, it is striking to consider how close the arc of human creativity has brought us. From shaping ink into stories to shaping cells into organs, we have crossed an invisible bridge between expressing thought and embodying it. As we walk further across that bridge, the responsibility to proceed with care and humanity becomes as important as the impulse to innovate.

Thought, once silent, now dares to breathe.

Epiphanies from This Wave

PART 4 – Health, Humanity and the Science of Life

— *A milkmaid's untouched skin whispered a secret: cowpox could guard against smallpox. From that observation came the miracle of vaccines.*

— *A drifting vapour lulled patients into sleep. With ether, pain lost its dominion, and surgery gained its mercy.*

— *A glowing shadow on a photographic plate revealed bones beneath flesh. Suddenly, the invisible body stood naked before science.*

— *Within the cell, a twisting staircase revealed life's instructions — a code that reshaped medicine forever.*

— *Printers once built toys and trinkets; now they shape organs that beat and breathe. What was play became the architecture of life.*

PART 5
Thoughts That Built Civilization
Economy, Shelter and Convenience

Chapter 1
The Thought of Trade – How We Began to Exchange the World

I imagine a time when there were no currencies, no receipts, no shipping invoices, and only need. A fisherman with more fish than he could eat meets a farmer with surplus grain. They lock eyes. A simple exchange takes place. No contracts, no coins, just mutual understanding. This was trade in its rawest form, barter shaped not by systems, but by survival.

Yet barter had limits. *What if* the farmer needed shoes, not fish? This dependency, often described as the double coincidence of wants, became the earliest barrier to exchange. It was here that the first real breakthrough appeared, not in goods, but in thought: the idea of money.

Around 600 BCE in Lydia, small stamped coins entered circulation. They did not spoil, did not need storage, and most importantly, people agreed they carried value. With currency came possibility. Trade expanded beyond local boundaries into unfamiliar lands. Paths became routes, and those routes gradually stitched civilizations together.

The Silk Road was not simply a corridor for silk or spices. It carried stories, languages, technologies, and ideas. Paper, gunpowder, and philosophical traditions moved alongside merchants and caravans. What began as commerce became cultural exchange. A trader moving between China and the Mediterranean unknowingly contributed to the foundations of a global economy.

Then came sails. Ships crossed oceans carrying sugar, coffee, and also human lives. The Age of Exploration expanded maps while narrowing moral boundaries. Commerce and colonialism grew side by side. It was a period marked by exploitation and suffering, yet it also accelerated global connectivity and economic interdependence.

Trade continued to evolve. The steam engine transformed speed, scale, and ambition. The Industrial Revolution reshaped production. Factories increased output, railroads crossed continents, and trade extended its reach with every new invention. Money evolved as well, flowing through banks, markets, and increasingly complex financial systems.

The twenty-first century introduced another shift. Trade became virtual, invisible, and constant. A click replaced a caravan. Markets no longer slept. A person could sit at home and purchase goods made halfway across the world, delivered in days or even hours. Platforms such as Amazon and Alibaba redefined commerce, while digital wallets carried value without physical form.

Yet even as trade advanced, new questions emerged. Can trade be ethical? Can economic growth coexist with environmental responsibility? Consumers began asking where their coffee was grown, how their clothes were made, and whether fair wages were paid. The trade of the future may depend not only on efficiency and profit, but also on accountability and values.

From fish exchanged for grain to one-click purchases, the story of trade mirrors the story of human imagination, desire, and responsibility. It began with a simple thought, and that thought continues to evolve.

Chapter 2
Money: From Barter to Blockchain

There was once a time when the world functioned without money. No coins caught the light. No paper notes exchanged hands. No numbers flickered across digital screens. There were only people, their needs, and the quiet logic of exchange. If you needed wheat, you offered eggs. If cloth was required, perhaps a goat or a basket of grain would suffice. Value existed in the present moment, shaped not by numbers but by usefulness. One person's need meets another's willingness.

Yet even within these simple exchanges, a deeper question began to surface. Could there be a more reliable way to trade? A shared language of value that could travel beyond immediate need, beyond place, beyond time?

When a Cow Was a Coin: The Age of Barter and Commodity

Barter was humanity's earliest economic system. It was direct, tangible, and deeply personal. For small communities, it worked well until it did not. What happened when the potter had no use for wheat? *What if* the farmer needed tools, not livestock? Trade stalled when desires failed to align.

To solve this friction, humans searched for something that could hold value independently of immediate exchange. Slowly, commodity money emerged. Cowry shells, salt, bronze, and cattle became trusted stand-ins for labor and resources. In many early societies, wealth was counted not in abstract units but in living assets, herds that could grow, move, and sustain life.

But these forms of money came with limits. They could rot, shatter, lose utility, or demand constant care. The idea of money had taken shape, but it had not yet reached efficiency.

When Trust Took Shape: The Birth of Coinage

Around 600 BCE, in the kingdom of Lydia, value gained a new form. Metal coins, stamped with official marks, entered circulation. Their worth came not only from gold or silver, but from authority. The seal of a ruler served as a public guarantee.

Coinage transformed trade. From Greece to Persia, from China to Rome, empires minted currency bearing symbols of power and identity. Money became standardized. Trust no longer depended on personal relationships, but on shared belief in the issuing authority.

Coins allowed trade to scale. Markets expanded. Economies connected strangers. Currency became a silent agreement between buyer and seller, state and citizen, past and future. Yet coins still carried weight, literally and figuratively. Humanity continued to search for something lighter.

When Paper Carried Value: The Thought Becomes Weightless

Centuries later, a radical shift occurred in China. During the Tang and Song dynasties, merchants began using paper notes. Value no longer required metal. A written promise was enough.

This innovation changed the pace of commerce. Wealth could move faster than caravans. Paper money spread across regions, carried by traders and institutions. In medieval Europe, promissory notes followed, eventually leading to official banknotes in Sweden and England.

Money no longer needed a physical substance to command trust. Belief, once inscribed, could travel vast distances. As money grew lighter, its psychological power increased. Value became increasingly abstract, yet no less real.

When Money Became Invisible: The Digital Shift

The twentieth century introduced another transformation. Plastic cards replaced cash. Credit, once a moral concept tied to reputation, became a routine financial tool. Transactions grew faster and less visible.

Then came the internet. Money dissolved into data. Transfers occurred through servers and signals. The sound of coins was replaced by confirmation tones and digital receipts. Currency left the physical world entirely, existing as entries in databases and encrypted systems.

Money had become invisible, yet omnipresent.

The Decentralized Dream: Cryptocurrency and a New Frontier

In 2009, a new idea entered the global conversation. Under the name Satoshi Nakamoto, Bitcoin was introduced as a form of money without centralized control. No bank. No mint. No governing authority. Trust shifted again, this time to mathematics.

Blockchain technology created a shared ledger, transparent and verifiable. Soon followed Ethereum, Litecoin, and state-backed digital currencies. Money evolved from a tool of institutions into an ongoing experiment in decentralization.

For the first time, money was not something you could physically possess. It was something you could verify.

Where Are We Headed

From barter to blockchain, the evolution of money mirrors the evolution of thought itself. Each stage reflects humanity's attempt to simplify exchange, strengthen trust, and expand cooperation.

These changes were never purely economic. They were philosophical. They shaped how societies understood value, ownership, and obligation.

Today, conversations turn toward cashless economies, decentralized finance, biometric payments, and AI-driven markets. A fingerprint may soon replace a wallet. A face may authorize a transaction.

Perhaps one day, money itself will disappear, leaving only value, exchanged through shared understanding rather than currency.

If that happens, money will have completed its longest journey. From substance to abstraction. From object to idea. From thought to thought.

Chapter 3
The Rise of the Financial Systems and Markets

I hold a coin and reflect on the power this small piece of metal possesses. Not because of its material composition, but because of what it signifies: trust, exchange, growth, and ambition. Within this modest object lies the resonance of a centuries-old idea, a thought that expanded into vast financial systems that quietly shape the rhythm of modern life.

Long before the word bank entered human language, people traded directly. A sack of grain for a goat. A tool in exchange for shelter. Yet as societies expanded and trade routes stretched across regions, barter revealed its limits. It was inefficient, inconsistent, and constrained by coincidence of need. Civilizations required something shared, dependable, and easily transferable. Money emerged as the solution. The Lydians of western Anatolia were among the first to mint standardized coins, embedding not only value, but consistency and trust into commercial exchange.

As money traveled, human thinking evolved alongside it. A new question emerged: what if value could move through time, not merely across space? *What if* one could give today and receive tomorrow? From this idea, credit was born. Clay tablets from ancient Sumer record debts and repayments, promises etched into hardened mud. In medieval Europe, promissory notes advanced the same principle, enabling merchants to trade across great distances without carrying physical wealth.

With credit came institutions designed to safeguard trust. In Babylon, temples functioned as repositories of wealth and issued loans. In ancient Egypt and Greece, early banking practices supported expanding trade networks. Yet it was Florence in the fourteenth century that transformed finance into a coherent system. The Medici family did more than operate a bank. They introduced transparency to financial practice. Double-entry bookkeeping created clear records of movement, ensuring that every transaction carried a balance and traceability. Money could now be observed, measured, and understood.

A more radical idea soon emerged. *What if* individuals could own a portion of a business itself? In 1602, the Dutch East India Company issued shares through the Amsterdam Stock Exchange, giving rise to the modern stock market. Capital no longer remained static. It could now be pooled, deployed, and multiplied. London and New York later adopted similar mechanisms, transforming public markets into powerful engines of innovation and expansion.

As markets expanded, the need for stability became paramount. In 1694, the Bank of England was established, widely regarded as the world's first central bank. Its function extended far beyond the safekeeping of funds. It regulated currency, managed interest rates, and reinforced confidence in the financial system. This institutional model spread across nations. Today, organizations such as the US Federal Reserve and the European Central Bank shape economic conditions through policy decisions whose effects ripple across entire societies.

Trust, however, has always remained fragile. Financial history is marked as much by collapse as by growth. The crash of 1929 and the financial crisis of 2008 exposed the dangers of speculation and unchecked ambition. Each upheaval prompted regulatory reform. Measures such as the Dodd-Frank Act were introduced to strengthen safeguards and reaffirm a central lesson of economic history: stability requires restraint.

In recent decades, money has shed much of its physical form. Paper yielded to digits. Payment services such as PayPal, Venmo, and UPI transformed transactions into seamless, nearly invisible exchanges. Blockchain technology and cryptocurrencies introduced a further possibility: decentralized systems operating beyond the authority of traditional banks. To some, this represents innovation. To others, it signals uncertainty.

Finance today is no longer confined within national borders. It functions as a global organism. A disruption in one economy can quickly affect another. Institutions such as the International Monetary Fund and the World Bank now play a role in maintaining balance across regions.

Looking ahead, financial systems continue to evolve. Artificial intelligence analyzes markets. Automated advisors guide investments. Central bank digital currencies may soon reshape how value is stored and transferred.

Yet beneath every system, tool, and transaction remains the same enduring question: how do humans value, share, risk, and trust? The coin in my hand is more than metal. It is memory, structure, and evolution. It is a thought that became a system, and a system that became a world.

Chapter 4
The Evolution of Banking: From a Thought to Civilization's Pillar

Money moves through the world much like blood moves through a living body. It circulates energy, sustains trade, allows cities to grow, and gives ideas the means to travel. Yet banking did not appear fully formed. No single thinker announced its arrival. It emerged gradually, shaped by commerce, necessity, and the fragile human capacity to trust. The history of banking is, at its core, the story of how civilizations learned to believe in promises and record them with care.

Clay Tablets and the Birth of Financial Thought

Before coins or paper, exchange relied on barter. A cow for a cart. Grain for oil. Over time, this system revealed its limits. Goods spoiled. Values conflicted. Fairness became difficult to measure. Around 3000 BCE, in the cities of Mesopotamia, the Sumerians introduced a quiet but revolutionary idea. They pressed symbols into wet clay, recording debts, credits, and transactions. These tablets allowed value to exist beyond the moment of exchange.

Temples became early vaults, storing grain, silver, and written records. Within their walls, a new form of thinking emerged. Wealth could be stored, loaned, and reclaimed later. This was not merely an economic shift, but a cognitive one. Value was no longer tied solely to physical goods. It could exist as an agreement, preserved in writing.

The World's First Bankers: Priests and Merchants

By 2000 BCE, Babylonian societies formalized these practices. Laws defined interest rates, collateral, and repayment obligations. These rules were etched into stone and clay, ensuring they would outlast individual traders. Priests initially supervised temple treasuries, but merchants soon expanded the system into marketplaces.

Clay receipts and promissory notes began to circulate. Money slowly transformed from metal and grain into meaning and trust. Banking evolved into more than storage. It became a structured belief system, one that relied on shared confidence rather than immediate exchange.

From Temples to Empires: Banking Grows Up

As trade expanded across Egypt, Greece, and Rome, banking evolved alongside it. In ancient Athens, money changers conducted business at trapeza, simple tables where coins were exchanged and loans negotiated. Rome developed more advanced institutions, including deposit houses and early payment drafts that reduced the need to transport large sums.

Centuries later, in medieval Italy, the banca, or wooden bench, became the symbol of financial exchange. When a banker failed, the bench was broken publicly, banca rotta. The word bankrupt still carries that memory. Language itself records the consequences of lost trust.

Rise, Fall, and Renewal

After the fall of Rome, formal banking systems fragmented. Trade declined, and with it, financial infrastructure. Yet necessity revived innovation. In the twelfth century, the Knights Templar created a network that allowed pilgrims to deposit funds in one region and withdraw them in another. This marked the emergence of international banking.

During the Renaissance, families like the Medici transformed finance into political and cultural power. Their development of double-entry bookkeeping provided a reliable method to track assets and liabilities. Banking shifted from serving trade to shaping it. Capital became a force capable of funding art, exploration, and empire.

From Ledgers to Digital Streams

By the eighteenth and nineteenth centuries, institutions such as the Bank of England and firms like J P Morgan stood at the center of industrial economies. Paper currency replaced metal. Credit expanded beyond physical reserves. In the twentieth century, electronic systems transformed banking again. ATMs, wire transfers, and digital ledgers accelerated the movement of value.

Today, cryptocurrencies and digital wallets exist in virtual space. Yet beneath algorithms and encryption remains the same ancient principle. Value is stored. Trust is exchanged.

Final Reflection: Banking is a Living Thought

From clay tablets to blockchains, banking has always been less about money than belief. It rests on the quiet assurance that tomorrow will honor today's promise. What began as simple marks in clay evolved into the invisible architecture of civilization. Banking remains a living idea, reshaped with every transaction that says, in essence, I trust you.

Chapter 5
The Evolution of the Share Market System

The story of the share market does not begin with glass towers or glowing tickers. It begins on wind-swept roads and crowded docks where merchants loaded caravans with spice, silk, and hope. Long before anyone spoke of stock exchanges, people were learning to share risk. A farmer might promise part of a harvest in return for a loan to fund a voyage. A shipowner could pledge future profits to those who supplied silver for trade. Trust itself was becoming a form of capital.

In ancient ports such as those of Mesopotamia and Phoenicia, contracts and partnerships spread risk across many hands. These arrangements were practical and modest, but they carried an important idea. Wealth could increase without goods having to move. The notion that value could be represented by a paper promise or a shared claim was taking root.

Venice and the First Paper Promises

By the twelfth century, Venice had become the mercantile heart of Europe. Its merchants issued bills and notes that represented credit and obligation. Governments borrowed against future taxes, and merchants used paper to finance voyages. These were not yet shared as we understand them, but they were the fragile beginnings of forms that would allow strangers to trust one another across distance.

The idea flowered into something fundamentally new in 1602. The Dutch East India Company offered investors small pieces of ownership in return for capital. For the first time, ordinary people could buy a share in a long-distance commercial enterprise. A sheet of parchment became a vessel of faith. Capital pooled from many investors allowed ships to sail farther, longer, and with greater resources than any single merchant could sustain. The company was not only a trading firm; it was an experiment in collective imagination.

The Age of Joint Stock and Speculation

The Dutch example inspired imitators. London and other ports adopted joint-stock ventures; the British East India Company and numerous private ventures created networks of investment that financed exploration, colonization, and war. Certificates and ledgers recorded ownership, but what really moved markets was belief. When many people believed that a venture would prosper, capital flowed, and enterprises expanded.

Speculation arrived with opportunity. Where trust gathered, rumor could travel, and fortunes were made and lost on whispers. The tide of confidence could lift whole markets, or sink them. That vulnerability became painfully clear in episodes of mania and crash, reminding investors that markets are not just machines of finance but stages for human emotion.

From Buttonwood to Wall Street

Across the Atlantic, a different chapter began beneath a buttonwood tree on a dusty New York street in 1792. Twenty-four brokers signed what became known as the Buttonwood Agreement, a compact to trade securities in an orderly way. From those modest roots grew the New York Stock Exchange. Wall Street, named for an old wooden palisade, would become the global artery of capital and risk. What began as a practical arrangement among traders evolved into an institution that channels savings into enterprise on a scale few early merchants could have imagined.

Industrialization and the Market as Imagination

During the nineteenth century, markets did more than redistribute goods. They financed an age of transformation. Investors bought shares in railroads that rebuilt continents, in steelworks that raised cities, and in electrical firms that illuminated night. Each purchase was a vote for a future. To buy a share was to bet that builders, engineers, and leaders could turn plans into reality. In that sense, markets were engines of imagination.

Crisis and Reform

Speculation can create astonishing growth, and it can also produce devastating collapse. The crash of 1929 exposed the fragility of systems built on unbridled confidence. In the aftermath, regulation and oversight emerged as necessary corrections. New laws, institutions, and practices aimed to stabilize markets and protect ordinary investors. The lesson was clear: markets without guardrails could become destructive.

The Electronic Age and the Speed of Money

The twentieth century introduced dramatic technological shifts. In 1971, the NASDAQ became the first electronic exchange, and over the following decades, trading moved from crowded floors to silent circuits. Algorithms replaced voices. Decisions that once took minutes or hours now occur in microseconds. The rise of electronic trading compressed time and amplified scale. Information moved faster than ever, and so did liquidity.

Today: Democratization and New Questions

Today, investing fits in a pocket. A retail trader can place orders on a smartphone app from anywhere in the world. Platforms such as Robinhood and similar services have made markets more accessible, bringing new participants and new perspectives. At the same time, technologies like blockchain and tokenization are challenging the very idea of ownership and clearing. They ask us what it means to hold value in a digital age.

This democratization brings both promise and risk. More people can participate, but speed and accessibility can fuel herd behavior. When millions act on the same signal, markets can swing wildly. The human heart still beats beneath the screens. Emotions, narratives, and collective belief remain the invisible forces that make markets alive.

Reflection: Imagination as Capital

From caravans to code, the share market has been less about money than about imagination. What began as simple agreements to share risk has grown into a global organism that channels collective faith into projects large and small. Every modern share traces its lineage to a basic, powerful idea: strangers can pool belief and capital to create something neither could do alone.

As we look forward, the market will continue to change. Regulation, technology, and social expectations will shape its contours. Yet underneath all change lies the same human impulse: the willingness to place a portion of trust and resources into a future not yet certain. That willingness is, in the end, the market's oldest currency.

Chapter 6
Wall Street's Rise from Wooden Wall to Financial Giant

It did not begin with towers of glass or men in dark suits. It began with a wall, a simple timber stockade built in 1653 when New Amsterdam was still a small Dutch outpost. The settlers raised the barrier to keep out threats they feared most: rival powers, the unknown of the wilderness, and any danger that might come from beyond. That wall was meant to protect a fragile settlement, not to create fortunes. Yet the street that later took its name would come to host the circulation of ideas and capital on a scale nobody then could imagine.

A dirt lane that once echoed with hooves and muskets gradually became a place where trust itself would be traded. By 1792, the wooden fortifications were long gone, and twenty-four brokers gathered beneath a buttonwood tree on a patch of ground off Wall Street. There, they signed what came to be called the Buttonwood Agreement, a compact to trade honestly and to give their markets order. No computers, no tickers, only voices and parchment. That modest pact planted the seed of the New York Stock Exchange.

Imagine that scene: men shaded by leaves, hands stained with ink, the sound of transactions, a simple exchange of trust. From this small, human beginning grew an ecosystem that would channel ambition into rails, mills, and factories. As the United States expanded, Wall Street financed the iron rails that connected coast to coast. It turned citizen-savers into underwriters of nation-building during the Civil War, when government bonds helped fund the Union's cause. Finance became an engine of expansion, the place where ideas could be translated into enormous material projects.

The story is not just about money. It is about imagination finding a market. What began as paper and handshake evolved into institutions: exchanges, brokerages, clearing houses, and later, regulatory bodies. The excesses of the 1920s, when credit flowed freely, and speculation intoxicated a generation, came to a brutal end on October 29, 1929, a day now remembered as Black Tuesday. Markets collapsed. Families lost savings. Confidence, the invisible currency underlying markets, was shaken.

From that collapse, reform followed. In 1933, the Glass-Steagall Act created walls of a different kind, separating commercial and investment banking to limit conflicts of interest. In 1934, the Securities and Exchange Commission was formed to oversee markets, protect investors, and bring a measure of transparency to trading. Regulation, imperfect and contested as it is, became part of the market's fabric because trust needed safeguards to survive.

After the Second World War, Wall Street shifted from servicing domestic ambitions to connecting a globalizing economy. American capital helped rebuild and then reshape the world. Multinational corporations rose. Consumer culture expanded. Finance did more than fund factories; it helped define modern aspiration. By the 1980s, the culture on the street had changed again. Deregulation and financial innovation accelerated activity and reward. Corporate raiders, leveraged buyouts, and the rise of complex instruments turned parts of finance into something like theater: dramatic, showy, and sometimes dangerous.

Today, the wooden lane exists mostly in name. The street has become a network of light and code. Trading is performed by algorithms and servers in temperature-controlled rooms. High-frequency trading executes orders in microseconds. Complex derivatives bundle risk into bundles that investors can buy and sell. The rhythm of markets is faster than ever, and yet the essential human impulse has not vanished. Behind every trade is a thought, a forecast, a question about the future. Every investment remains, at its heart, an act of confidence.

That faith is fragile. The financial crash of 2008 revealed how quickly systems that seem robust can unravel when incentives are misaligned and risk is misunderstood. Excess had once again outpaced caution. In response, reform and oversight returned, and the debate about regulation versus innovation resumed with renewed intensity. Each crisis has a lesson: markets adapt, but they learn through pain.

The digital age introduces new paradoxes. Technology has increased efficiency and lowered costs, but it has also concentrated power and speed. Algorithms do not feel; they act. That removes some human error, yet it introduces risks we do not always foresee. Blockchain and cryptocurrencies promise alternative structures for trust, distributed ledgers that do not need a single gatekeeper. They are experiments in rethinking what money can be and who controls it.

If you walk down Wall Street today, the sensory world is different from that of 1792. You will see glass, steel, and digital ticker displays. You will hear less shouting and more quiet clicks in server rooms. Yet the deeper pattern is familiar: human imagination finding new ways to allocate resources and to make collective decisions about the future.

When I reflect on this history, I feel both admiration and caution. Wall Street's rise is a testament to the capacity of organized thought to fuel industry, science, and collective projects. It shows how shared belief can build railways, hospitals, and universities. But it also teaches humility. Power without reflection can become habit and harm. Profit without purpose can hollow out societies.

The street that once divided to protect a colony now connects economies across the globe. From piled timber to digital ledgers, Wall Street's story is a mirror of modern life: full of invention, capable of great creation, and always in need of governance and moral clarity. Its legacy is neither wholly heroic nor wholly tragic. It is a human story, complex and unfinished, one in which trust must be rebuilt again and again, and where imagination must be wedded to responsibility if the market is to serve more than itself.

Chapter 7
Humanity's Ascent from Shelter to Skyline

Once, shelter meant a roof, a wall, and a fire. It was a small set of promises: keep out the cold, hide from rain, offer a place to sleep. Early dwellings were local and literal, built from mud, straw, timber, or stone. They answered a simple need and, by doing so, created a quiet ritual of safety. Even at that scale, a home felt like a small island against an unpredictable world.

Across millennia, those islands began to gather. Villages became towns. Towns became cities. And with urban growth came a new question: how high could we go?

Ancient builders reached up for reasons that were often spiritual as much as practical. In Mesopotamia, the Sumerians stacked mud and brick into ziggurats, stepped monuments meant to bring the heavens closer to the gods. In Egypt, pyramids rose not to house the living but to secure the dead against time, each stone laid as an argument with mortality. The Romans, more worldly in their aims, built multi-story apartment blocks for dense urban life. Some Roman insulae reached as high as nine stories by the first century BCE, a risky but practical response to crowded streets.

These early climbs were limited by material and method. Walls had to be thick to bear weight. Open space was expensive. Fire and collapse were constant threats. For centuries, height remained a gesture rather than a practical city strategy.

A spark in the furnace: steel and elevators

The modern ability to build tall depends on two inventions working together: a new structural material and a practical way for people to reach higher floors.

In the mid-19th century, steelmaking changed everything. Henry Bessemer's method, developed in the 1850s, allowed manufacturers to produce strong, consistent steel at scale. Suddenly, builders had a material with a high strength-to-weight ratio, one that could form a skeletal frame instead of relying on thick load-bearing walls. Steel frames meant more glass, more light, larger interiors, and taller structures.

Equally important was vertical transport. In 1853, Elisha Otis demonstrated a safety elevator that prevented cars from falling if the hoisting cable broke. That demonstration transformed fear into confidence. A building could now be tall because people could move through it safely and quickly. Steel and elevators together rewired what architects and developers imagined.

The Home Insurance Building in Chicago, completed in 1885, is often called the first skyscraper because it used a structural steel frame to support its ten stories. It was a modest beginning compared with today's towers, but its method was decisive. The building announced a new logic: height could be systematic rather than accidental.

Foundations, form, and the birth of the skyline

At the turn of the 20th century, Chicago and New York became laboratories for vertical architecture. Architects and engineers explored how form could follow function, and how aesthetics might arise from engineering. Louis Sullivan captured this idea in a phrase that would shape modern architecture: form follows function. He meant that the design should emerge from a building's purpose rather than from applied ornament alone.

Skyscrapers became civic statements as well as functional solutions. The Woolworth Building, completed in 1913, rose to 792 feet and announced the commercial optimism of its era. The Chrysler Building, completed in 1930, added an Art Deco crown that read like a triumph in metal. In 1931, the Empire State Building reached a symbolic height, 1,454 feet to its tip, and for decades it stood as a monument to ambition and technical achievement.

These towers did more than puncture the skyline. They reordered urban life. They concentrated offices, commerce, and services into vertical neighborhoods. They changed how people moved, worked, and imagined the city itself.

Skyline today: new materials, new purposes

The appetite for altitude has only grown. Late 20th and early 21st century towers blend cultural iconography, engineering innovation, and environmental thinking. The Petronas Towers, completed in 1998, used Islamic geometric principles to shape a linked twin tower silhouette. The Burj Khalifa, completed in 2010, reached 2,717 feet and became a study in how scale, structure, and landscape interact.

At the same time, architects and engineers are asking different questions. How should tall buildings live with the environment? Can towers produce their own energy, reduce their water footprint, and support biodiversity on their terraces? Concepts such as vertical gardens, solar skin, and wind-harvesting facades are moving from experiment to practice. Plans for even taller structures, like the proposed Jeddah Tower, have been ambitious, though timelines and outcomes may shift with finance and politics.

From shelter to skyline: a human story

What I find most revealing is how the shift from shelter to skyline reflects a change in human expectation. Early homes answered the need. Modern towers articulate identity, economy, and aspiration. They are tools of density and symbols of possibility. They compress the city into vertical blocks where office life, homes, culture, and transport must coexist in tighter quarters.

There are costs and tradeoffs. Tall buildings can cast long shadows over neighborhoods, strain infrastructure, and require complex safety systems. They can concentrate wealth as well as opportunity. That is why design matters: form must respect context, and engineering must be married to civic responsibility.

A brief reflection

When I walk beneath a tall building, I feel a mix of admiration and humility. There is drama in a skyline, but there is also humility in the labor that made it possible: the steelworkers who braved heights, the engineers who solved unexpected problems, the craftspeople who finished intricate interiors. A skyline is a long conversation between imagination and constraint.

The movement from simple shelters to soaring towers is one of humanity's most visible stories. It shows how materials, technology, and social organization can reshape the places we live. It also shows how an answer to a practical problem can become a cultural statement. If you stand in a modern city at dusk and watch the lights come on floor by floor, you are witnessing a history of questions and small solutions stacked into an architecture of aspiration.

We are not finished with the skyline. As climate concerns and new materials reshape our priorities, the next generation of towers will likely look and behave very differently. Yet the impulse behind them will be familiar: humans imagining upward, inventing a way to climb, and then sharing the view.

Chapter 8
The Thought of Convenience: Measuring Time

Time was not always a number. For long stretches of human history, it flowed in the slow work of sun and season, in the lengthening shadow across a field, in the hush that follows dusk. People lived by rhythms they felt in their bodies and read in the sky. But at some point, a quieter question took hold: could we shape time so it could shape us? Could we make its passing useful, shared, and certain?

When the sky was our first clock

Long before gears or glass, the heavens kept watch. The sun's rise and set marked beginnings and endings. The moon's cycle guided planting and prayer. The stars described patient arcs that people learned to read, to remember, to trust. Around the second millennium BCE, sundials began to appear. Stones cut to catch sunlight recorded the hour as a shadow moved, turning the day into a simple, visible count.

Night required different tools. Water clocks, or clepsydras, measure time by the steady fall of liquid. Every drop was a quiet pulse. In some cultures, such vessels were placed in tombs as a promise that time itself might follow the dead. These devices did not make time absolute, but they made it shareable. For the first time, moments could be compared across distance, and people could begin to coordinate beyond immediate sight.

When time took shape in gears

As societies grew larger and more complex, the need for a common time became urgent. Markets, religious rites, and civic life all depended on people arriving together. Mechanical clocks began to appear in Europe in the late medieval era, often housed in church towers where their sound could reach a whole town. The clock in Salisbury Cathedral, built around 1386, began to mark public life in a new way. Its swinging weights and tolling bells made time a civic rhythm rather than a private feeling.

With the Industrial Revolution, the scale of movement changed everything. Trains ran on rails that ignored the sun. Each town, once content to set noon by its own sky, suddenly found that a single day contained many local noons. Schedules clashed, and accidents were not rare. The problem was practical and pressing. A Canadian engineer, Sandford Fleming, proposed a bold remedy. He suggested dividing the globe into twenty-four standard time zones. In 1884, at an international conference in Washington, nations accepted Greenwich as the prime meridian, and the modern framework of coordinated time took shape. Railways, telegraphs, and the flows of commerce now ran to the same pulse.

When time became personal

For centuries, public clocks ruled the day. Then technology placed time in the pocket. The portable watch, refined in the sixteenth century, allowed individuals to carry moments with them. It was more than a tool; it was a symbol of order and autonomy.

The wristwatch reached broad use for practical reasons in the early twentieth century. Soldiers in World War I needed a quick, reliable way to coordinate maneuvers without fumbling for a pocket watch. A strap transformed a private instrument into a wearable device. After the war, the wristwatch remained. It became part of a daily ritual: a glance, a plan, a small act of control.

When quartz silenced the ticking

Accuracy moved from artisan hands to invisible vibration. In 1969, Seiko introduced the first mass-produced quartz watch. Instead of gears and escapements, a tiny crystal oscillated at a precise frequency when driven by electricity. The result was a quiet, nearly perfect timekeeper that lost only seconds per year. The romantic tick of mechanical clocks faded in many places, and with it a certain affection for audible time. Yet what replaced it was extraordinary reliability. The world could now depend on instruments that kept an almost flawless beat.

When atoms began to count

Today, the clocks that govern global systems are often invisible. Atomic clocks measure time by the vibrations of atoms, usually cesium or rubidium. These oscillations are so regular that modern atomic clocks lose only a second in millions of years.

They synchronize satellite navigation, power grids, and the global internet. Their precision allows planes to cross continents safely and financial markets to record transactions to fractions of a second. The pulse of modern life depends on clocks that most of us never see.

The thought that ticked forward

Each advance in timekeeping answered a different human need. Sundials bound communities to the sky. Water clocks and tower bells organized markets and prayer. Pocket watches marked status and independence. Military necessity turned the watch into a companion on the wrist. Quartz made accuracy ordinary. Atomic clocks made synchronization global.

But these inventions are not merely mechanical achievements. They reflect how people relate to one another and to their own lives. Measuring time answered a deeper longing: the wish to place our fleeting days into a shared order, to plan and promise, to meet and remember together. As devices grew more precise, they also made new demands. The same instruments that keep trains on schedule now allow systems to trade, coordinate, and govern at speeds that challenge human judgment.

A brief caution

With better clocks came new questions. When everything is timed and tracked, who decides which moments matter? What does it mean when calendars, devices, and servers shape our attention and our work? The tools that make life orderly can also make it measured and compressed. We gain efficiency, but we risk losing some kinds of leisure that come from loose hours and idle afternoons.

When I consider the arc from shadow to atom, I feel a mixture of admiration and humility. There is wonder in a stone that holds sunlight and in a device that counts the heartbeat of an atom. These inventions show the human will to shape the world, to find rhythm in chaos. They also remind me how much our communal lives depend on conventions we rarely question.

Time is a human invention that became the stage for so many other inventions. It is not simply a thing we measure. It is the scaffolding of work and love, of markets and memory. From carved sundials to invisible atomic pulses, the story of timekeeping is a story of us trying to live together.

What remains constant is the thought that began it all: a simple question asked long ago, perhaps while watching a dawn or bending to plant a seed. *What if* we could hold time, even if only a little, so that our lives might find common measure and meaning?

Time moves on, and thought moves with it. Between the sun's shadow and the atom's vibration, we keep asking that oldest question: what shall we do with the hours we are given?

Chapter 9
Glass – The Thought That Captured Light

There is something quiet and uncanny about glass. It holds water without showing seams, it reflects sunlight while remaining almost invisible, it separates inside from outside yet lets the world pour through in light. Glass feels eternal, but it is the product of many human decisions, experiments, and accidents. It is an idea that learned to melt sand and, in doing so, changed how we live.

Born of Fire, Sand, and Curiosity

The first glass was not manufactured. It arrived in moments of violent heat. Lightning striking sand and volcanic flows that cooled into obsidian gave early people a glimpse of what hardened, glossy minerals could look like. Those natural glasses suggested a possibility: what if humans could make that gloss at will?

By around 3500 BCE, craftsmen in Mesopotamia were fashioning beads from a lustrous, glasslike substance. In Egypt, artisans were making colored glass for amulets, jars, and inlays. These early objects were rare and symbolic. They were worn and traded as if they contained something of the sacred. For centuries, glass remained a wonder, admired more than used.

From Ornament to Everyday Object

The transformation began with technique. In the first century BCE in the eastern Mediterranean, glassblowing emerged and spread quickly. By shaping molten material with breath and a tool, makers could produce vessels with speed and variety that had been impossible before. Suddenly, glass left the temples and entered the marketplace. Cups and flasks moved from luxury into domestic life.

This was not simply a technical change. It altered how people related to objects. Glass turned private rituals into ordinary acts. A window lets light into a room without smoke or a draft. A simple cup makes drinking less brittle and more graceful.

The material that had once been admired in small beads now became a quiet companion of daily life.

Machines, Medicine, and Mass Production

Industrialization made glass ordinary in a new sense. Factories learned to press and mold it, to float it into broad sheets, and to produce it in quantities that stretched across continents. By the nineteenth century, glass was a staple of kitchens, laboratories, and factories. It was cheap enough for common use and precise enough for science.

The laboratory needs modern medicine to shape glass into a tool of trust. Beakers, vials, syringes, and microscopes rely on predictable, inert glass. The transparency of glass allowed scientists to observe experiments directly. In hospitals, sterile glass became part of a system designed to see and to safeguard. In optics, lenses made from glass expanded human sight. Telescopes and microscopes both depend on glass to bring the distant and the small into focus.

The Modern Age: Everywhere and Nowhere

Walk through a modern home, and you will find glass in countless, often invisible roles. Double-glazed windows regulate temperature, mirrors shape self-image, and screens translate light into the interfaces we use to work and communicate. Fiber optics carry our voices and images across oceans, using thin threads of glass to guide light with astonishing efficiency. The screens in our pockets are engineered glass, toughened and polished to resist the daily wear of human life.

Glass is made by turning silica sand, soda ash, and lime into a molten mass at very high temperatures, typically around fourteen hundred to sixteen hundred degrees Celsius, then cooling it with care. The process is both art and engineering. Small changes in composition or cooling speed produce very different materials, from the delicate clarity of crystal to the flexible strength of tempered glass.

A Thought Still Unfolding

The story is not finished. New forms of glass are emerging that respond to light or temperature, that change opacity with a signal, or that carry vast amounts of data at the speed of light. Photonic chips, microstructured glass for medical devices, and advanced lenses for astronomy extend what the material can do. In medicine, glass continues to enable diagnostics and delivery systems that were unthinkable a generation ago.

When I think about glass, I am struck by how a handful of natural events and a string of practical experiments produced something so central to modern life. It began with curiosity about a glossy stone and became a technology that frames our days and shelters our labs. Glass quietly mediates the world. It lets us see more clearly, but it also asks us to consider the choices behind what we now take for granted.

Glass began as a thought: what if a solid could be transparent? That question opened a path from beads and amulets to fiber optics and medical imaging. It illustrates how a simple curiosity, tested and refined over centuries, can become the scaffolding of civilization. Glass does not shout. It reveals. And in its reveal, it has changed the way we know the world.

Chapter 10
The Thought That Changed Food: Refrigeration

There was a time when every kitchen kept an eye on the weather and the clock. Milk curdled in the heat before breakfast. The meat darkened by evening. Ice was a treasure, chipped from lakes and guarded as if it were gold. People learned to eat with the season and to plan with care. Then someone asked a simple, radical question: *What if* we could catch a cold and keep it? That question changed how we stocked our pantries, how we fed cities, and how we traded across oceans.

Borrowing cold from the world

Before machines, people borrowed cold from the world around them. In deserts, water cooled in porous clay jars as the wind drew away heat. In cooler regions, communities harvested winter ice and stored it under straw and earth to last into summer. These are small, human inventions that feel intimate: a child leaning over a jar to feel the chill, a merchant supervising blocks of ice loaded with sawdust. The cold was valuable because it was scarce. Finding ways to hold it became a kind of measured wisdom.

Creating cold on purpose

The move from harvesting cold to making it was slow and leapwise. Early experiments in thermodynamics showed that cooling could be produced by physical means. Inventors pushed that idea into practical machines. Engineers such as Jacob Perkins and James Harrison refined mechanical refrigeration into devices that could keep breweries and meatpacking houses chilled. A physician in the nineteenth century imagined cool air as therapy and built a primitive cooling apparatus to ease fevers. These efforts were not only technical; they were moral in purpose. People wanted to preserve life and reduce suffering.

The cold that traveled

Before refrigeration filled homes, ice itself became a trade. Merchants cut and shipped frozen blocks around the world. They wrapped ice in insulating materials and sent it where warmth ruled. This trade connected climates and cuisines. It allowed fresh foods to travel across far distances for the first time. Then came refrigerated rail cars, a breakthrough that changed where cities could obtain fresh meat and produce. Cold no longer belonged only to winter or to places with mountain springs. It began to move along rail lines and shipping lanes, redefining diets and markets.

Cooling the household

The real revolution occurred when cold moved into the home. Early iceboxes required regular deliveries, a daily choreography between household and supplier. Electric refrigeration shifted that balance quietly and profoundly. An appliance humming in the kitchen meant vegetables lasted longer, children could drink fresh milk without fear, and households wasted less food. Refrigeration changed how families shopped, how restaurants planned menus, and how small businesses survived.

Shadows in the wake of progress

Progress carried costs. The chemicals that made early refrigerators efficient, especially chlorofluorocarbons, had an unseen effect high in the atmosphere. Scientists discovered that some of these compounds damaged the ozone layer, which protects life from ultraviolet radiation. The discovery prompted an international response. Nations agreed to phase out the most harmful chemicals, and engineers developed safer refrigerants and new designs. That moment reminded us that invention must be coupled with responsibility. When we master one part of nature, we must watch how that mastery touches other parts.

The next wave: smarter, cleaner cold

Refrigeration keeps evolving. New technologies offer cooling without the old environmental cost. Magnetic refrigeration and advanced compression systems reduce harmful emissions. Solar-powered coolers supply vaccines and food to places without steady electricity, protecting lives where a grid does not reach. Smart systems, guided by simple algorithms or by artificial intelligence, adjust temperatures, predict demand, and cut waste. Refrigeration has become not only a tool for keeping food safe but an instrument for managing resources more wisely.

A quiet revolution that sustains life

From porous clay to electric compressors, the story of refrigeration is a story of human patience and practical imagination. It is an argument that small conveniences can carry vast consequences. By learning to hold cold, we stretched the season, expanded trade, and protected health. We also learned a lesson about limits: the technologies we embrace change the web of life around us. That is why the work continues, not only to keep food fresh, but to keep the planet livable.

When I think of refrigeration, I think of a childhood kitchen where the simple presence of cool air meant peace of mind. That feeling is not trivial. It is the felt consequence of many curious minds and patient hands. Great inventions often begin with modest aims: to keep milk from souring, to stop meat from spoiling, to spare a fevered patient needless suffering. Those aims ripple outward until they alter economies and ecosystems.

If refrigeration began with a question that seemed small and domestic. And later grew into a global force that touched trade, health, and the climate. The lesson it leaves us is practical and urgent. New inventions must be paired with foresight. We must measure not only what our machines can do, but what the world can bear. If we keep that balance, the cold we create will continue to sustain life without costing us the future.

Chapter 11
Wrapped in Thought – A Quiet Journey Through Packaging

Early mornings have a particular stillness. I sometimes feel it most when I open a parcel. There is a soft crackle as plastic shifts, a faint scent rising from whatever waits inside. The moment carries a kind of ceremony. A box is not only a container. It is a promise that something once distant has made its way safely to my hands. As I tear the paper, I often find myself thinking about how this simple act connects us to a very old idea: the instinct to protect what matters.

Packaging is more than cardboard, tape, and branding. It is the story of how human beings learned to preserve what they valued, transport what they needed, and share what they created. From the natural shells that guarded seeds to the algorithms that now design shipping routes, the evolution of packaging has traveled quietly beside our own.

Nature, the First Wrapper

Long before cities existed, and long before language could describe protection, nature was already demonstrating it. Seeds are hidden inside hard shells. Fruit protected its sweetness beneath thick skin. A cocoon held a fragile life with astonishing precision. Early humans simply watched and learned.

Coconut husks became bowls. Banana leaves wrapped food. Animal skins stored water. These choices were not about aesthetics. They were about survival. Food needed to travel without spoiling.

Water needed to remain clean. Fire and insects needed to be kept out. Packaging began as an instinct, not an industry.

By roughly three thousand years before the common era, civilizations in Egypt and Mesopotamia were sealing grains, oils, and medicines in clay jars and sealing them with wax or mud. These vessels were not simply for storage. They were tools of endurance, protecting essential goods from heat, pests, and time.

Paper, Reeds, and the Rise of Craft

As societies grew, so did their ability to refine materials. In ancient China, around one hundred years before the common era, soft sheets of mulberry bark were used to wrap herbs, spices, and delicate items. Egypt developed papyrus and woven reed containers, transforming everyday storage into a form of craftsmanship.

Across India, Asia, and the Mediterranean, woven baskets of straw, bamboo, and cane carried grains and fruit. These packaging methods were lightweight and renewable. In many ways, they were more environmentally responsible than the materials that dominate modern commerce.

The Romans introduced durability on a new scale. Their wooden barrels, crates, and clay amphorae allowed wine and olive oil to cross oceans intact. Packaging became a companion to trade, enabling goods to travel farther than the maker ever would.

The Industrial Age and the Wrapping of the World

With the arrival of steamships and railroads, the world began to shrink. Food and goods had to travel greater distances and last longer. In 1810, the tin was transformed into preservation. A sealed metal container meant food could survive months or even years without decay.

By the late nineteenth century, cardboard and corrugated paper reshaped commerce. These materials were light, inexpensive, and strong enough to protect fragile items. Glass jars arrived for sauces, preserves, and medicines. Printed labels became the first layer of branding, telling stories before the product inside was even opened.

Companies like Heinz and Nestle realized that packaging could influence trust as much as taste. A well-made container signaled quality. A clean seal reassured families. Packaging stepped beyond protection and entered the realm of persuasion.

Plastic: The Great Shift

In 1907, the invention of Bakelite marked the birth of fully synthetic plastic. It was heat-resistant, moldable, and durable. But the real transformation arrived in the 1930s with polyethylene, a flexible and transparent material that soon touched every corner of modern life.

After the Second World War, plastic became the skin of consumer society. Bottles, trays, films, wrapping, caps: everything became lighter, cheaper, and astonishingly resilient.

Convenience soared. But something else happened. Plastic refused to disappear.

This durability, once celebrated, revealed its darker edge. Plastic accumulated in rivers and oceans. It broke down into particles too small to see yet too persistent to remove. The miracle material lingered long after its purpose was fulfilled. Humanity had mastered the art of wrapping, but now found itself wrapped in the consequences.

The Convenience That Stayed Too Long

Convenience often hides a cost. Plastic packaging, designed to protect goods, began to threaten ecosystems instead. Microplastics drifted through water, soil, and air. Wildlife ingested fragments that would never degrade. Waste sites grew into mountains.

These effects pushed scientists, engineers, and designers to rethink the purpose of packaging. If the problem came from human invention, then the solution must come from human intention.

The Future: Smarter, Softer, Kinder

Today, innovation is guided by conscience as much as convenience. New materials aim to protect both the product and the planet.

Seaweed-based films dissolve into soil without harm. Mushroom-based packaging replaces foam. Bioplastics from sugarcane and cassava break down within months rather than centuries. Edible wrappers hint at a future where waste no longer shadows consumption.

Technology deepens this transformation. Sensors inside bottles track freshness. Smart labels monitor temperature. Packaging communicates with refrigerators and supply chains. Artificial intelligence designs boxes that use less material and stack more efficiently, reducing cost and emissions.

The measure of good packaging is changing. It is no longer about lasting forever. It is about leaving a gentle footprint.

From Leaves to Algorithms

From fish wrapped in broad leaves to electronics sealed in multilayered foils, packaging has always reflected the needs and values of the societies that create it. It shows our ingenuity and our blind spots. It tracks our progression from survival to abundance and, now, toward responsibility.

As I hold a simple parcel in my hands, I sometimes feel the weight of that long history. Packaging is more than a protective layer. It is a record of how we have learned to care for the things we make and the world we live in. And perhaps the next chapter is not only about designing stronger wrappers, but about designing wiser ones.

The best packaging does more than protect what is inside. It respects everything outside.

Chapter 12
The Blade That Rewrote the Morning

Every morning begins with small rituals. One of them is older than industry, older than nations, older than the idea of comfort itself. A hand reaches out, closes around a familiar handle, and prepares for a brief moment of precision. Most of us never think about the history behind this simple act. Yet the story of how we shave traces a long human journey shaped by discomfort, necessity, invention, and a single question that quietly changed modern life.

From Stone to Steel: The First Cut

Long before razors came packaged in plastic or powered by electricity, early humans carved edges from stone. Flint and obsidian offered sharpness for a brief moment before their brittle ends dulled. Shaving was not about appearance. It was a practical matter. A heavy beard could harbor insects, trap heat, or interfere with battle gear. Removing hair meant protection and comfort.

By around 3,000 BCE, Egyptians crafted copper and bronze blades and elevated shaving into a ritual. Cleanliness expressed purity, and priests kept their bodies entirely shaven. Later, the Romans introduced iron blades and turned shaving into a social act. Barbers became part craftsmen and part storytellers, trimming beards while political debates echoed through stone bathhouses. A shave remained a careful task. One slip could draw blood.

The Straight Razor: A Tool of Skill and Trust

By the eighteenth century, the straight razor became the tool of choice. Forged from high-quality steel and sharpened on leather, it was elegant and dangerous. Skilled hands handled it with pride. Men gathered in barbershops, where the morning sun caught the shine of metal and conversations drifted between jokes, anxieties, and public news. Shaving required time and concentration. It was not rushed. It was a ritual.

Still, a persistent inconvenience hummed in the background. Blades dulled quickly. Sharpening was unavoidable. The process was slow, and mistakes were painful. The ritual held beauty, but it also held frustration.

King Camp Gillette: The Man Who Thought Differently

In 1895, King Camp Gillette, a traveling salesman, stood before a mirror in Boston and faced the nuisance familiar to every man of his time. His straight razor was dull again. He was late again. And the routine had worn thin. Gillette did not respond with anger but with imagination. Why must a blade demand so much work? Why could it not simply be replaced?

Disposable products were beginning to emerge. In his job at Crown Cork and Seal, Gillette saw bottle caps used once and discarded. He noticed patterns. Convenience was reshaping manufacturing. He wondered whether the razor could follow the same path.

His idea was simple. A thin blade made cheaply and replaced easily. No sharpening. No fear of slipping. No need for a trained hand.

Many dismissed him. Metallurgists claimed steel could not be manufactured that thin and remain strong. Investors doubted the market. But Gillette kept the idea alive until he found an engineer who could match his belief.

From Idea to Industry: Building the Impossible

William Emery Nickerson, trained at MIT, saw potential where others saw impossibility. Together, he and Gillette built machines capable of stamping ultra-thin blades with consistency. What had once been a frustration became a product, and in 1901, they founded the Gillette Safety Razor Company.

The early numbers were small. In its first year, the company sold just over fifty razors. Yet the world was shifting toward convenience. By 1904, nearly ninety thousand razors and millions of blades were sold. The transformation had begun quietly, almost invisibly.

A private irritation had become an industry.

The War That Made the World Shave

Then came World War I. In the trenches, hygiene and speed were survival. A clean shave allowed soldiers to secure gas masks fully. Gillette approached the U.S. military with a proposal. Every soldier would receive a safety razor as part of his standard kit.

Young men learned to shave themselves, quickly and safely, without relying on barbers. When they returned home, the habit returned with them. What had been military discipline became civilian identity. Morning routines changed across the country almost overnight.

A Blade, a Business Model, and a Blueprint

Gillette's genius extended beyond the blade. He pioneered a business model that would later shape entire industries. The razor would be affordable. The blades, replaced repeatedly, would generate lasting revenue. This approach became the blueprint for printers and ink, coffee machines and pods, video game consoles and software.

Gillette did not simply redefine shaving. He helped redefine consumption.

Sharper, Safer, Smarter: Innovation Returns Again and Again

Shaving continued to evolve. Stainless steel blades appeared in the 1950s, lasting longer and resisting rust. Multi-blade systems arrived in the 1970s, seeking smoother results. Handles improved. Heads pivoted. Lubrication strips softened the edge. Electric shavers offered speed and convenience. Modern companies have adopted subscription models, turning blades into a monthly service delivered to doorsteps.

Yet beneath each new feature lies Gillette's original thought. A tool should solve a problem without demanding unnecessary effort. It should be simple, clean, and reliable.

The Thought That Endures

King Camp Gillette is remembered not because he crafted metal better than anyone but because he reimagined a daily routine with clarity. He turned a moment of irritation into an idea that spread across continents. He saw complexity and chose simplicity. He believed that even a small tool can shape culture.

Each morning when a blade touches skin, a quiet echo of that thought appears. It is a reminder that innovation often begins with a single question asked at the right moment. Could this be easier?

Sometimes the most ordinary habits hold extraordinary stories. And sometimes one sharp idea is enough to change how the world begins its day.

Chapter 13
When Sound Learned to Stay

Long before technology learned how to store sound, artists depended entirely on human memory to preserve their work.

There was a time when music lived only in the air. A note was born, it trembled for a moment, and then it vanished. Every song was fragile. Every performance is irreversible. Around fires, in temples, in village squares, people sang not to record, but to remember. Melody was sacred precisely because it could not be held. Once it faded, it survived only in the mind.

Then a quiet, world-changing question emerged. *What if* sound could stay?

Capturing the Invisible: Edison and the First Recorded Voice

In 1877, inside Thomas Edison's cluttered laboratory, a small metal cylinder wrapped in tin foil began to turn. As it rotated, sound vibrations pressed into the surface, and something extraordinary happened. When the cylinder was played back, a human voice emerged.

For the first time in history, sound had been trapped.

Edison called his invention the phonograph. It was not just a machine. It was a new relationship between humanity and time. A voice could now outlive the body that created it. A moment no longer had to disappear. The ephemeral form had been given form.

Within years, recorded sound moved from curiosity to utility. Speeches, songs, and stories could be preserved and repeated. Memory had become mechanical. The human voice had gained the ability to travel forward through time.

From that spinning cylinder arose a radical truth. What once vanished into the night could now echo across generations.

From Furniture to Palm: The Shrinking of Sound

As the decades passed, the phonograph evolved into the gramophone. Its large, horn-filled rooms are filled with amplified sound. Music moved from concert halls into private homes. It became furniture, an object that occupied space.

Then came the transistor.

By the mid-twentieth century, the radio no longer had to sit in one place. It could travel. Kitchens, cars, streets, and fields filled with sound. Music followed daily life instead of waiting for an audience to approach it.

Soon after, the cassette arrived. Small. Magnetic. Transformative.

For the first time, listeners were not just consumers. They became editors of their own history. Mixtapes were assembled with intention. Love, loss, joy, rebellion, all arranged in sequence. Technology had not only preserved sound. It had handed authorship to ordinary people.

Music was no longer only performed. It was curated.

Freedom in Headphones: The Walkman Era

In the late 1970s, Masaru Ibuka of Sony asked a simple question. Why can't I take music wherever I go?

The answer became the Walkman.

With it came a revolution in listening. Headphones created private worlds in public spaces. Streets gained soundtracks. Commutes turned cinematic. For the first time, music did not belong to the room. It belonged to the individual.

This was not just portability. It was psychological freedom. A person could move through the world accompanied by a chosen emotional atmosphere. Music became an identity in motion.

It was not merely a device. It was independence.

The iPod and the Digitization of Memory

In 2001, Steve Jobs posed another quietly powerful question. *What if* you could carry a thousand songs in your pocket?

The iPod was not the first digital music player. But it was the first to make complexity disappear behind elegance. Paired with iTunes, it shifted music from physical objects to a digital experience. Shelves no longer bowed under the weight of albums. Songs became invisible files arranged by mood, memory, and meaning.

Ownership was no longer tied to plastic or paper. It was tied to access and organization.

Music had compressed itself into light.

Listening was no longer a ritual. It became a seamless extension of daily life.

When the Air Itself Began to Sing

By the late 2000s, ownership began to dissolve altogether.

First came Napster, unsettling the industry. Then Spotify, Apple Music, and streaming platforms remade the idea of access. The entire history of recorded sound became available instantly. Music stopped living on shelves or devices. It lived in the cloud.

We stopped holding music and began living inside it.

A song could find us before we searched for it. A playlist could mirror our emotions before we name them. Sound became ambient, like oxygen. Invisible but everywhere.

Humanity has turned music into an atmosphere.

The Infinite Listener: AI and the Age of Prediction

Today, we whisper to machines, and they answer with melody. Artificial intelligence does not simply play songs. It predicts desire. It learns patterns, habits, and moods. It shapes soundtracks for our inner lives.

Music now exists primarily as data. We no longer own it. We access it. We breathe it.

Yet beneath this seamless convenience lies the same ancient longing that first moved Edison to listen differently. The need to preserve what moves us.

The desire to keep feeling alive after the moment has passed.

We not only wanted the sound to stay. We wanted the memory to stay.

Closing Thought

This is not merely the history of devices. It is the story of how thought learned to listen.

It began with a question filled with wonder. Can sound remain? Can memory be held? Can emotion echo beyond its origin?

From a tin cylinder to invisible streams of data, every invention has whispered the same answer in a new voice.

Yes, sound can stay. Memory can be heard.

And through it all, human thought continues to hum, persistent, searching, and beautifully alive.

Chapter 14
The Silent Force Behind Civilization – A Thought That Moved the World

We often celebrate generals, rulers, and inventors, but the quieter idea behind their power is hardly ever praised: how do we move things? Not armies alone, but food, water, stone, grain, timber, and steel. The pulse of civilization has always depended on motion, the steady, unglamorous work of supply. Long before the word logistics existed, people were solving this problem in mud and sweat, by river and road, because without reliable movement, nothing else endures.

When I click Buy Now, and a parcel arrives at my door within hours, I am not only enjoying convenience. I am standing on the shoulders of millennia. Every delivery echoes camel caravans that crossed deserts, sailors who trusted the wind, and minds who turned movement into mastery. This is not only the story of transport. It is how civilization learned to keep itself alive.

The Earliest Movers, Written in Dust and Sweat

Thousands of years ago, logistics was not a theory; it was a fact of survival. In Egypt, around 3000 BCE, workers hauled massive stone blocks across hot plains without cranes or engines. Building a pyramid required routes planned with care, teams timed to pull and rest, and supplies moved along the Nile with such precision that the monument became not just a religious work but a proof of coordination. Every rope pull, every baked-brick tally, was an act of collective engineering.

In Mesopotamia, the same lesson appears in clay. Tablet after tablet recorded grain, livestock, and trade. The Sumerians did not invent bureaucracy for its own sake; they learned a practical truth: progress demands order. You must know what you have, where it is, and how to move it. From those inventories grew accounting, and from accounting came the first maps of supply.

Across continents and centuries, the Silk Road stitched empires together.

It was not a single path but a network of pulses, caravans carrying silk and spices, ideas and risk. The route was rough and slow, yet its value was not merely commerce. It connected distant peoples in ways many modern systems still struggle to match. Before logistics had a name, it had a heartbeat.

Logistics Marches with Armies

War taught planners a harsh clarity: an army cannot live on courage alone. The Roman legions are famous for their roads, but those roads were instruments of supply. Grain, oil, tools, and tents moved in disciplined order. Portable bridges and field forges let armies operate far from home because the logistics chain made a sustained presence possible.

Napoleon expressed this in a blunt, practical line: an army marches on its stomach. When his supply lines froze in the Russian winter, his forces collapsed. The lesson was not new, but its consequences were stark. Logistics had moved from the background to a decisive factor. Victory depended not only on bravery but on reliable delivery.

Steam, Steel, and the Tracks of Change

Then steam arrived, and with it, speed. Railways rewired continents, and steamships made oceans more predictable. The sound of engines replaced the silence of sails. Ports came alive, cities swelled, and goods began to move with a steadiness the world had never known. Distance collapsed, not through imagination, but through iron and fire.

Inside factories, motion itself became choreography. Henry Ford is often credited with inventing the assembly line, but his true genius lay elsewhere. He did not create the idea; he transformed it. Inspired by the overhead conveyor systems of meatpacking plants and the moving belts of grain mills, Ford reimagined manufacturing by bringing the work to the worker. In 1913, with the first moving assembly line for the Model T, he broke production into dozens of small, precise steps and set the product itself in motion.

The result was revolutionary. A car that once took more than twelve hours to build could now be assembled in less than two. Costs fell. Access widened. What had been a luxury became attainable. Efficiency was no longer an abstract principle; it became the currency of progress, reshaping labor, time, and industrial ambition.

War soon revealed the deeper power of logistics. The twentieth century made movement visible on a scale never before seen. World War II was won not only by strategy and courage, but by supply lines and coordination. The Red Ball Express, a relentless chain of trucks surging across a shattered Europe, kept front-line troops fed with fuel, ammunition, and food. Steel, rubber, and gasoline moved like lifeblood through the arteries of war. The side that moved faster, cleaner, and more reliably often held the advantage.

Logistics had become destiny.

Barcodes and Data: The Invisible Grid Forms

In the 1970s, the revolution turned quiet again. A pattern of black lines on a label began to speak. The barcode lets a product announce its presence with a single scan. Warehouses learned to track inventory as a live flow rather than a monthly accounting exercise. Computers replaced ledgers, and supply chains became measurable networks.

Companies like FedEx and UPS turned delivery into an exact science. They measured time by the minute and space by the second. Knowing where something was becoming as important as moving it. With each digital beep, logistics crossed from visible routes to a hidden grid that binds the globe.

Click, Buy, Fly: The Age of E-Commerce

Then the internet changed expectations. Speed moved from desirable to demanded. Customers wanted not just fast delivery, but immediate fulfillment. Amazon and other platforms reorganized warehouses, automated processes, and used algorithms to predict demand. Same-day shipping stopped being a rarity and became a baseline.

Behind a single click, an orchestra plays: robots pack orders, algorithms route trucks, humans solve exceptions. The supply chain stopped being a line and became a living web. The world wrapped itself with a network designed to answer the question of movement in ever-smaller time windows.

AI at the Helm: The Future Already Moving

Today, logistics does not merely react; it predicts. Artificial intelligence anticipates demand, reroutes shipments mid-journey to avoid delays, and balances energy use across entire fleets. Drones lift lightly to remote doorsteps. Autonomous trucks traverse highways at night. Warehouse robots learn to find the fastest path to a pallet. Sustainability has entered the logic of electric vans, and optimized routes reduce carbon footprints. The old question, how to move from here to there, has become sharper: how do we move faster, cleaner, and smarter while preserving the human dignity behind every motion?

The Unseen Thought That Moves Us All

From camel caravan to drone, logistics has always been more than machines. It is a thought that reaches across time: how do we create reliable movement for the needs of many? Every bridge, every route, every supply ledger is an answer to that question. Civilizations rose because someone found ways to move grain to the hungry and stone to the builders. The daily miracle of a parcel arriving at a door is the same impulse that fed armies and built cities.

When I study this history, I feel a quiet gratitude. The designers of roads and the clerks who counted grain rarely receive applause, but their work knit societies together. They taught us that progress is not only invention but systems, small decisions, and steady practices that transform possibility into routine.

Ethics and Limits

Seeing the machinery behind civilization also brings responsibility. Speed and efficiency have costs. Labor can be undervalued; ecosystems can be stressed; communities can become dependent on fragile supply chains. As logistics grows smarter, we must ask not only how to move more, but for whom and at what cost. The power to deliver can also concentrate power in the hands of a few. The choice then becomes one of governance and care.

The thought of movement has matured. It now asks broader questions: how do we balance speed with justice, and convenience with resilience? How do we make systems that serve people and preserve the planet?

A Final Thought

Logistics is a silent force behind every advance we celebrate. It is the patient architecture that turns ideas into everyday life. Civilizations are not built by ideas alone, but by the systems that carry them forward. Every delivery, every bridge, every ledger is part of the same story: the effort to move what matters from where it is to where it must be.

Perhaps the greatest innovation of all is recognizing that movement is itself a form of care. When we move food to a table, medicine to a clinic, and books to a child, we are doing more than transporting goods. We are extending the possibility. We are, in a practical manner, making civilization possible.

Chapter 15
When Shopping Became a Quiet Walk-Through Thought

I step into a supermarket.

Cool air settles on my skin, steady and familiar. The lights stretch in neat rows above me, each one casting a soft glow over shelves arranged with almost mathematical precision. Colors spill outward in every direction: the greens of vegetables, the silver of tins, the warm tones of bread and grain. Somewhere in the background, a scanner releases its brief electronic chirp, a steady reminder of movement and exchange. The automatic door closes behind me without a word.

I push my cart forward. The wheels glide across the polished floor with a sound so soft it almost disappears. Picking up a carton of milk or choosing a piece of fruit feels effortless, almost instinctive. And yet a quiet question lingers just beneath the surface of all this familiarity.

Was it always this simple?

What Looks Ordinary Was Once Revolutionary

Shopping today feels calm and unhurried. No clerk stands at the counter to ask what I need. No negotiations. No balancing of metal weights on a scale. I walk freely, choosing what I want with my own hands. A single beep at the register seals the exchange.

But this frictionless experience was once unimaginable. Not long ago, shops were small and intimate. The customer asked; the shopkeeper delivered. No self-selection. No wandering. No comparing. Someone, at some moment, looked at that system and wondered:

What if people could choose for themselves?

What if everything they needed lived in one place?

That question became the seed of a revolution.

When Shopping Meant Knowing Your Butcher

Not long ago, shopping was personal.

The baker greeted you by name. The butcher remembered the exact cut your family preferred. The grocer trusted you enough to grant credit. Each interaction carried a human warmth; a kind of micro-community built across counters and conversations.

But intimacy came at a cost.

Queues grew. Inventory was limited. Prices were higher. As cities expanded, the pace of life quickened. People needed something more efficient, something that matched the growing rhythm of modern society. A quiet new thought waited, ready to reshape everything.

The First Shift

Clarence Saunders and the Pig That Made People Think

In 1916, in Memphis, Tennessee, a grocer named Clarence Saunders looked at the traditional store and saw a puzzle waiting for a solution. He wondered what would happen if customers could pick out goods themselves. No clerks standing between the shelves and the shopper. Just aisles, baskets, and the freedom to choose.

He called his store Piggly Wiggly. The name made people smile, but the idea made them think.

Inside his store, customers walked independently through the aisles. They compared brands, checked prices, and examined produce. At the end of the trip, they brought their choices to a single checkout counter.

It was more than efficient.

It felt empowering.

For the first time, ordinary people became active participants in the act of shopping.

The Big Leap: Michael Cullen and the Market Made Super

A few years later, Michael Cullen expanded the idea. In 1930, in Queens, New York, he opened King Kullen. This store was unlike anything people had seen. It was large, bright, and stocked with an enormous variety of goods. It had wide aisles and large parking spaces for the growing number of families who now owned cars.

Cullen called it a supermarket. The name fits.

It was super in scale, super in selection, and super in its ambition.

Here, shopping became more than a transaction. It became an experience, a small ritual of modern life.

The Innovations That Followed

Once the door opened, progress flowed steadily.

In 1937, the first shopping carts rolled into stores, allowing shoppers to collect more goods with ease.

In 1974, a barcode on a pack of Wrigley's gum became the first product scanned at a checkout counter. That small beep marked the beginning of an invisible revolution. From then on, data moved faster, errors decreased, and queues shortened.

Freezers and refrigerators expanded the reach of the modern diet. Foods that would spoil within hours could now last for days or weeks. Frozen peas, ice cream, and packaged meals became fixtures of the weekly household routine.

Checkout belts moved products smoothly. Air conditioning cooled the store through long summers. Soft music filled the aisles. What had once been a small shop now functioned as a finely tuned machine built for comfort and speed.

Every improvement began as a single idea.

Small, simple, but powerful.

From Supermarkets to Hypermarkets

By the 1960s and 1970s, a new question surfaced:

What if everything lived under one enormous roof?

The answer arrived in the form of hypermarkets. Carrefour in France and Walmart in the United States built vast spaces where groceries, clothing, electronics, and household goods existed side by side. Shoppers could complete an entire week of errands in a single trip.

Scale created affordability.

Affordability created loyalty.

And loyalty reshaped the global economy.

Shopping became a weekend routine. Families pushed carts down fluorescent aisles, moving through a landscape designed for both necessity and possibility.

Then the Walls Disappeared

A new shift came quietly. The internet arrived.

Suddenly, you did not need aisles or carts. There were no queues, no closing hours, and no distance between the shopper and the store. Online shopping systems like Amazon, Instacart, and others were built directly upon the same question that Saunders once asked.

How can this be easier?

A tap replaced the touch.

A click replaced the conversation.

Delivery replaced the drive.

The supermarket went digital, becoming both everywhere and invisible. Yet its purpose remained the same: to give time back to the mind that created it.

What the Supermarket Really Reflects

When we walk through a supermarket, we see oranges, pasta, detergent, bread, and packaged goods. But if we look carefully, we see something deeper. We see the evolution of convenience. We see a transition from community-based trade to a system designed around individual freedom and efficiency.

We no longer trade words across counters.

We trade glances with barcodes.

The supermarket is not simply a building. It is an idea. A physical expression of society's desire to organize chaos into clarity.

Every automatic door, every scanner's beep, every neatly aligned shelf speaks to a quieter truth:

Progress begins with a question.

Not a loud question, but a small one, asked with sincerity.

How can life be easier?

How can life be better?

Behind every seemingly effortless act of shopping are decades of inventions, failures, insights, and successes. And within the hum of machines and the steady rhythm of movement, one can still hear the quiet pulse of thought, walking alongside us through every aisle.

Chapter 16
A Thought in a Cup: How Coffee Stirred the World Awake

One sip, and the mind lifts. Not with noise, but with a new clarity, a subtle sharpening that changes how the day begins. Coffee is ordinary enough to be overlooked and extraordinary enough to remake routines, cities, and conversations. Behind that steaming cup lies a long, surprising history of curiosity, exchange, and human connection.

Life before coffee

Imagine a morning without that impulse to reach for something hot and bitter. Before coffee spread across the world, daily life followed the rhythm of daylight. Wakefulness arrived with the sun, and tiredness returned with darkness. Work and thought moved at the pace of the body. There was patience in that pace, but there was also a limit to how far reason and debate could extend into the night.

A goat herder noticed something odd

Stories say a shepherd observed his herd behaving strangely after eating the berries of a wild shrub. Their energy and alertness caught attention. A few humans tried the berries too and felt a new wakefulness, a shift in attention that could be used for long watches and long prayers. Whether the tale is literal or a tidy origin myth, the simple fact matters: at some point, people discovered beans that change the tempo of the mind.

From the highlands to the shared table

By the fifteenth century, cultivation and trade moved coffee across regions and cultures. In Yemeni towns, mystics used the brew to sustain hours of meditation and night prayer. Coffeehouses began to appear, places where people gathered to exchange news, argue, recite poetry, and listen. The drink became an aid to concentration and a catalyst for public life. It was not merely a stimulant; it created social spaces where minds could meet.

Coffeehouses as public forums

When coffeehouses reached European cities in the seventeenth century, they altered the culture of conversation. In London and Paris, these rooms of talk became hubs of commerce and learning. For the price of a small coin, anyone could enter, hear the news, meet a merchant, or listen to a visiting scholar. People spoke across social lines in ways that had been rare in private salons or closed societies. These were places where ideas circulated as readily as coins.

From suspicion to acceptance

The early arrival of coffee in Europe was met with suspicion by some religious authorities who worried about its effects on morals and order. Yet curiosity and taste prevailed. A celebrated anecdote tells of religious leaders tasting the drink and finding it agreeable; once approved, coffee moved more freely into daily life. The beverage that had accompanied devotion across oceans now warmed philosophers, statesmen, and artisans.

Fuel for industry and thought

The arrival of the Industrial Revolution changed both machines and the rhythms of labor. Coffee moved from the margins to the center of a modern workday. In factories and offices, the break for coffee became a familiar pause that sharpened attention, recharged hands, and created brief rituals of shared conversation. Steam engines and coffee breaks moved in parallel: one powered industry, the other energized the minds that designed and ran it.

Instant coffee and modern convenience

In the twentieth century, innovations condensed the ritual into convenience. Instant coffee made it possible to stir energy wherever a kettle and hot water existed. Soldiers carried it into war zones; travelers kept it for long journeys; households stored it on pantry shelves. Companies grew around the bean and the brand, transforming a local habit into a global market. A small cup connected coffee farmers on distant hills to office workers in skyscrapers, linking supply and demand across continents.

A global habit with local variants

Today, coffee appears in countless forms and local customs. Specialty roasters, neighborhood cafés, and multinational chains all offer different versions of the same basic idea: a brewed infusion that alters attention. For some, a single cup begins the work of the day. For others, the ritual of sharing a drink brings people together, creating conversation that can lead to collaboration or to quiet companionship. Coffee is practical and symbolic at once.

Economic and ethical reflections

Coffee's global reach has consequences. It is among the most traded commodities on Earth. That scale creates livelihoods for millions, but it also raises questions about fairness, sustainability, and labor. The modern coffee trade has prompted innovations in farming, export, and certification, and it has forced consumers to consider the origins of their daily habit. The cup in your hand links you to a long supply chain and to workers whose labor makes that moment possible.

A thought that changed civilization

What does a cup of coffee do to a culture? It extends attention, it lengthens the day, and it produces spaces where ideas can be exchanged. Coffeehouses incubated discussions that helped shape politics, literature, and commerce. In that way, the beverage became a tool of public life, an enhancer of thought, a liquid that lubricated debate.

A quiet, human note

When I think about these histories, I feel a small, private gratitude. A physicist's experiment, a trader's ship, a farmer's careful harvest, a barista's steady hand, and a writer's late night all meet in a single cup. That continuity fascinates me. It reminds me that great changes often begin with ordinary acts, repeated across time and place. One sip connects you to a chain of human choices that makes longer, more attentive days possible.

An everyday invitation

As you lift your cup, consider this: beneath the steam and the warmth lies a human story of curiosity, commerce, ritual, and work. Coffee did not create genius, but it changed how we sustained attention, how we shared ideas, and how we organized public life. It is both a humble comfort and a social technology, one that helped turn small conversations into movements.

So when you take another sip, remember you are part of that lineage. That warm liquid in your hand is less a simple drink and more a partnership between chemistry and human will. One cup, one conversation, one idea at a time.

Chapter 17
A Leaf That Stirred the World: The Story of Tea

Across the highlands of ancient Yunnan, where morning mist moves like slow breath across the hills, a single leaf once drifted into a pot of boiling water. The story blends fact and legend, but its impact is unmistakable. A man lifted the cup, expecting nothing more than warmth. Instead, he discovered a calm clarity that would travel farther than armies, trade routes, or kings.

Some revolutions burst into existence with noise. Others unfold quietly. Tea belonged to the second kind. A simple sip began a global habit, a moment of stillness that shaped cultures, economies, and empires.

A Chance Encounter: The Birth of a Thoughtful Brew

The earliest tale speaks of Emperor Shen Nong around 2737 BCE, often described as a curious experimenter who believed that boiling water protected against illness. Resting beneath a camellia tree, he noticed leaves falling into his pot. The resulting drink surprised him. It sharpened his senses while steadying his mind. Whether this event happened exactly as told is uncertain, yet its symbolic truth remains. Tea became one of the earliest examples of humans discovering alertness without agitation, peace without sleep.

A leaf had introduced a new state of mind. That first sip invited reflection. It encouraged a pause before action. It became a companion to thought itself.

Tea as Medicine: Before the Ritual Came the Remedy

Long before tea became a beverage for pleasure, it served as medicine. Early Chinese texts, including the Shennong Ben Cao Jing, described it as a purifier that refreshed the body and cleared the mind. It was seen not only as a drink but as a remedy for fatigue, fever, and inflammation.

Buddhist monks later embraced it for a different purpose. In the quiet rhythm of monasteries, tea helped maintain wakefulness during long hours of meditation.

It supported discipline without disturbing inner balance. In these halls, a warm cup replaced the need for drums or bells. Awareness rose with every sip.

By the Tang Dynasty, tea had grown beyond medicine. It became culture. Poet and scholar Lu Yu wrote The Classic of Tea, codifying everything from cultivation to brewing. His work treated tea as a practice that merged precision, observation, and respect. A beverage became an art. A routine became a philosophy.

Tea Travels: From China to the World

For centuries, China remained the heart of tea culture. Leaves were pressed into cakes for transport, their value treated with the seriousness of fine silk. As caravans crossed the Silk Road, tea spread slowly outward.

When Buddhist monks carried seeds to Japan, something deeper happened. Tea found not just a new geography but a new spiritual shape. By the early ninth century, the monk Eichu planted the first tea bushes near Kyoto. Over time, Japanese tea culture evolved into chanoyu, the tea ceremony. It was not merely a drink but a deliberate sequence of gestures. Every motion from bowing to whisking became a meditation on harmony, respect, and awareness.

In Japan, tea was not consumed. It was an experience.

The Empire and the Leaf: Britain's Brewing Obsession

Tea arrived in Europe in the seventeenth century, carried by ships of the Dutch and later the British East India Company. It fascinated traders and aristocrats alike. In Britain, it quickly became a luxury, then a necessity.

By the eighteenth century, tea was central not only to daily life but to politics. The 1773 Boston Tea Party remains one of the most symbolic acts in American history. Colonists were not simply rejecting taxation. They were declaring that everyday rituals could carry the weight of rebellion. A cup became a catalyst.

Unable to access China's production secrets, Britain turned to its colonies. In the hills of Assam and Darjeeling, vast plantations began to reshape global commerce. The story included conflict and exploitation as well. Botanist Robert Fortune discreetly transported tea plants and knowledge from China to India, a bold act that shifted the balance of trade. Tea was no longer just a beverage. It was a force that shaped economies.

The Many Faces of Tea

Wherever tea traveled, it adapted.

In Britain, it became afternoon tea, introduced by the Duchess of Bedford, turning a simple break into a social ritual.

In Russia, the glowing samovar warmed homes, producing a drink that moved slowly through hours of conversation.

In India, chai became part of the country's pulse. On railway platforms, kettles hissed and vendors called out to travelers as milk, black tea, and spices merged in familiar comfort.

In Morocco, mint tea reflected hospitality, poured from high above to release scent and sweetness.

In China, the traditional gongfu cha continued to honor precision and patience, using tiny cups to intensify flavor and attention.

Each region shaped tea into its own philosophy. Some made it communal. Some made it reflective. All made it meaningful.

Industrial Sips: From Loose Leaf to Tea Bags

For centuries, brewing tea demanded time and focus. Water had to be heated carefully. Leaves had to open slowly. But the twentieth century valued speed. In 1908, a New York merchant named Thomas Sullivan sent tea samples in small silk bags. Customers dipped them straight into hot water, misunderstanding the packaging as a brewing tool. That accident created the modern tea bag.

Tea changed. It became faster and more convenient. Industrial processing replaced traditional methods in many places. Yet tradition persisted, too. Some people defended the teapot as a quiet rebellion against the pace of modern life.

Iced tea also emerged during this era. At the 1904 St. Louis World's Fair, Richard Blechynden offered chilled tea to cope with the heat. America embraced it, and a warm-weather staple was born.

The Wellness Wave: Tea for Body and Soul

As the twentieth century ended and the twenty-first began, society returned to tea with new intentions. Research into antioxidants renewed global interest in green tea. Herbal blends gained popularity as gentle remedies. Cafes and tea houses offered slow spaces in fast cities.

Matcha became a favorite among those seeking focus. Jasmine pearls and oolong teas displayed the artistry of rolling, drying, and scenting leaves. People began treating tea not just as a drink but as a ritual of balance in a world filled with noise.

Brewing Intelligence: When Tea Meets Technology

Innovation eventually reached even the tea fields. Artificial intelligence now analyzes soil chemistry, predicts harvesting windows, and tracks fermentation. Blockchain allows consumers to trace the journey of each batch from mountain to market. Smart kettles adjust temperature automatically, learning preferences over time.

Despite these advancements, the essence of tea remains unchanged. It is still a conversation between water, leaf, and heat. Technology may refine the process, but it cannot replace the quiet pause that tea invites.

Final Sip: The Leaf That Anchored a Billion Thoughts

From the legendary moment under a camellia tree to the global networks of today, tea has remained a companion to human reflection. It has survived empires, revolutions, migrations, and inventions. Through all of history's noise, it continues to offer a moment of stillness.

Every cup carries the same message whispered by that first drifting leaf:

Slow down. Sip. Think.

It is a reminder that even in a world driven by speed and innovation, some of our most enduring comforts begin with a pause.

Chapter 18
How One Advertisement Redefined Santa for Generations

Santa Claus did not arrive fully formed. For centuries, he wandered through stories and villages, wearing many faces. In some places, he appeared as a solemn gift-bringer in fur and brown, in others as a kindly bishop in flowing white robes. He was a composite of saints, sailors, winter spirits, and storytellers, a figure shaped by the needs of local imagination rather than by any single portrait.

That began to change in the early twentieth century, when commerce and culture converged beneath electric holiday lights. In the depths of winter, soda sales cooled with the weather, and Coca-Cola wanted a different winter story, one that could warm both hands and hearts. The company commissioned Haddon Sundblom to create a series of images that would feel like a heart in a bottle, familiar and comforting to readers of magazines and passersby on city streets.

Sundblom did not invent Santa from nothing. He looked to an already beloved poem, A Visit from St. Nicholas, first published in 1823, which had fixed many details of the modern myth: the sleigh, the reindeer, the playful entry through chimneys. Sundblom gave that poem a human face. He painted a man who might sit and listen, whose mouth could surprise a child with a low laugh, whose eyes held private mischief. He dressed him in warm garments colored to match Coca-Cola's own palette of red and white, and he treated him as a neighbor rather than a saint.

The images were simple in composition and rich in feeling. A child's cheek warmed by lamp light. A coat textured by brush strokes.

A mug cradled in mittened hands. Sundblom's Santa was tactile. He smelled of wood smoke and sugar; he felt like a story told aloud on a cold night. Between 1931 and 1964, these paintings ran in magazines, posters, and storefront windows, appearing year after year until the sight of that round, rosy face became familiar in homes across the country and beyond.

Coca-Cola did not invent the idea of Santa Claus, but the campaign crystallized a particular image and made it nearly universal. A thousand local versions of a winter spirit, each with its own color and costume, gradually folded into one widely shared figure: jolly, red-suited, generous. Advertising gave the older folklore a consistent silhouette that could travel with print and film, and later with television.

The modern Santa is an example of how a culture can borrow an image, polish it, and then propagate it until it feels like memory. It also shows how business and myth can shape each other. A single advertising campaign did not create Christmas, but it did reorganize how millions pictured their holiday. For many, the man Sundblom painted is the man who comes down the chimney.

There is a quiet lesson here about influence and taste. Stories arrive layered, then get hauled into the light by poets, painters, and merchants. What survives is rarely the oldest version. It is whatever image was repeated until it could not be forgotten. For a hundred million households, the red-suited, laughing figure is that image, a shared shorthand for warmth, generosity, and the season itself.

When I think of Sundblom's work, I feel a mixture of admiration and curiosity. He did not flatten a myth; he honed one of its many faces with care. In doing so, he helped give people a way to imagine generosity that was at once intimate and public. That is a subtle power: the ability to shape how a culture pictures its own kindness.

Chapter 19
A Thought That Split the World

It did not begin with thunder.

There was no flash. No roar. No rising pillar of smoke.

It began in a quiet laboratory, beneath steady lamps and steady minds. A question formed, fragile and almost innocent. Could the matter we touch, the metal in our hands, the stone beneath our feet, carry hidden forces vast enough to reshape the world? Deep inside the atom, could there be energy bound tighter than any known power?

At first, this was not a weapon. It was a curiosity. A hunger to understand what the universe is truly made of.

The slow opening of the atom

The story unfolded over decades, not in a single moment. Marie Curie isolated new elements and revealed that matter itself could radiate unseen energy. Ernest Rutherford probed the atom's inner structure and showed that it was not solid but mostly empty space. Niels Bohr gave form to its strange internal order through mathematics. Each discovery pulled back one more veil.

Then, in 1938, Lise Meitner, working with Otto Hahn and building on earlier experiments, interpreted a result that changed physics forever. The uranium atom had split. Energy was released, not as a theory, but as a measurable fact. Soon after, Otto Frisch confirmed the process and named it nuclear fission. For the first time, humanity understood that the atom itself could be broken, and that in breaking it, immense energy would be freed.

In a quiet winter laboratory, the universe had revealed one of its deepest mechanisms.

From theory to controlled fire

Understanding is one thing. Control is another.

In 1942, under the stands of a squash court at the University of Chicago, Enrico Fermi and his team assembled the world's first nuclear reactor, Chicago Pile One. Layer by careful layer, they built a structure of graphite, uranium, and precise mathematics. There was no explosion when the switch was thrown. There was only a subtle rise in activity, a steady and self-sustaining chain reaction held in human hands for the first time.

The invisible had become controllable. Thought had turned into usable energy.

At that moment, nuclear power stopped being an abstract idea. It became a tool.

The brighter promise of the atom

In its peaceful form, nuclear energy offered something rare in human history: vast power without smoke. A small amount of fuel could generate enormous electricity. Unlike coal or oil, it did not fill the sky with soot or carbon dioxide. It worked without sunlight and without wind. It produced heat day and night, steady and consistent.

Nuclear methods also transformed medicine. Radiation therapy emerged as a way to shrink tumors. Nuclear imaging began to trace disease within the body. Reactors produced isotopes that saved lives daily in hospitals around the world.

For a time, it seemed as though humanity had found a near-perfect source of energy, one that could support growing cities and ambitious science without darkening the skies.

The shadow that followed

But the atom revealed two spirits.

Even as reactors were being built for light and medicine, the same physics was being shaped into the most destructive weapons ever created. In August 1945, atomic bombs were dropped on Hiroshima and Nagasaki. In seconds, entire sections of cities were erased. Tens of thousands died instantly.

Many more suffered in the years that followed. The power once glimpsed in laboratory equations had now torn through human lives.

Later came industrial disasters that deepened the fear. At Chernobyl in 1986, a flawed reactor design and human error released radioactive material across large parts of Europe. In 2011, a massive earthquake and tsunami led to reactor failures at Fukushima. These events did not destroy the technology, but they burned into public memory the cost of mistakes.

Nuclear waste added another layer of unease. Some byproducts remain dangerous for thousands of years. The problem of safe storage does not belong to one generation. It stretches far beyond it.

The moral divide

Nuclear thought sits at a rare crossroads of human conscience.

It lights cities and powers hospitals. It also carries the memory of cities erased in seconds. It offers carbon-free energy at scale, but it demands absolute care. No other technology asks so fiercely for precision while punishing failure so severely.

The tension is not merely technical. It is ethical. How should such power be governed? Who should decide its limits? How do we ensure that knowledge meant to serve does not destroy?

These questions do not fade with time. They deepen.

Returning to the atom with humility

Today, a new generation of scientists and engineers approaches nuclear power more cautiously. Safer reactor designs are being tested. Small modular reactors aim to reduce risk. Fusion research, still experimental, seeks energy from the same process that fuels the stars without the long-lived waste of fission. International treaties work to limit the spread of weapons, even as their presence continues to shadow global politics.

The thought that once split the world still lives at this crossroads. It carries both warning and promise.

A reflection

When I study the history of the atom, I feel both awe and unease. Few ideas have traveled so swiftly from curiosity to a civilization-shaping force. Fewer still have revealed so clearly how a single thought can divide our future into paths of light and darkness.

Nuclear science was never born as a weapon. It began as a desire to understand what lies beneath the surface of reality. Yet once the door was opened, humanity had to choose how to walk through it.

Perhaps that is the real lesson of this chapter. Knowledge itself is rarely evil or pure. It is the hand that holds it that defines the outcome. The thought that split the atom also split our responsibility. And that responsibility has not ended. It grows with every generation that inherits its power.

Chapter 20
The Man Most Guarded – A Glimpse into the World's Highest Protection

Power wears many faces. One is public, arranged for cameras and crowds. Another is invisible, operating without applause to keep a single person alive so that an entire nation can continue to function. At the center of that invisible world stands the President of the United States. He moves through space surrounded by layers of planning, technology, intelligence, and people who have turned vigilance into a lifelong discipline. What the public notices is the wave, the speech, the hand on the podium. What the public rarely sees is the immense system quietly working to prevent catastrophe before it ever reaches the light.

The House and What Lies Beneath

The White House stands on Pennsylvania Avenue like a familiar painting, a structure so endlessly photographed that its deeper purpose can fade into the background. Visitors see wide lawns, historic columns, and roses in geometric order. Yet beneath that calm surface lies a hardened world of reinforced corridors, secure chambers, and protected systems designed to function when everything else fails.

Deep below the West Wing sits the Presidential Emergency Operations Center. It exists for moments no leader hopes to face. It is built to preserve command, protect decision makers, and maintain communication if the world above fractures. It is not a symbol. It is a contingency made of steel, data, and silence.

Above ground, another layer of defense remains constantly alert. Cameras interpret movement. Sensors measure intrusion. Access systems judge identity in fractions of a second. Across Washington, restricted airspace and radar frameworks sweep the skies. These measures are not ceremonial. They are responses shaped by history, by attacks that forced painful learning, and by the unyielding truth that the consequences of failure are beyond calculation.

To sustain this invisible shield, vast resources are required. In fiscal year 2024 alone, the United States Secret Service operated on a budget of approximately 2.96 billion dollars. That figure is not merely an accounting line. It represents personnel, intelligence operations, cyber defense, transport security, emergency medicine, and the constant evolution of counter threats in a world where danger no longer announces itself in advance.

Protection at this level is not built once and maintained forever. It is rebuilt each day through analysis, rehearsal, surveillance, and acceptance among those who serve; their success is measured by what never happens.

An organization born of necessity

The agency most associated with personal protection is the U.S. Secret Service. It began in 1865 to fight counterfeit currency. The job changed when an assassin's bullet ended President McKinley's life in 1901. After that, protection became its primary calling.

Since then, the agency has grown into a specialized force. Agents train in tactics and medical care, in threat assessment, and in anticipation. They study crowds until patterns look obvious to them and invisible to the rest of us. They plan routes, rehearse responses, and imagine what could go wrong before it ever does.

The Beast - U.S. Presidential State Car loaded onto a U.S. Air Force transport aircraft.

When the president steps into public life on the road, he is surrounded by a convoy that functions like a moving fortress. The central vehicle, commonly called The Beast, looks like a limousine, but inside it is a sealed environment built to survive explosions, biological agents, and gunfire. Tires keep rolling after damage. Windows are thick and sealed. Communications connect directly to military and intelligence networks. Medical supplies, including a supply matched to the president's blood type, are on board.

The motorcade is far from spontaneous. Each route is surveyed in advance. Roads can be closed, manhole covers welded, and contingencies rehearsed for every junction. The choreography includes decoy vehicles, medical teams, counter-assault units, and communications platforms. Every movement is timed and redundant.

The visible calm of a motorcade masks hours and days of preparation.

Air travel is protection in another dimension. Air Force One is more than an aircraft. It is a flying command post with hardened systems to protect from electromagnetic disruption and nuclear effects. It can refuel in flight, maintain secure communications, and become a government hub when necessary. For shorter hops, the president takes Marine One, a helicopter flight wrapped in formation flying that hides which aircraft carries the president. The precision of those maneuvers is part practicality and part performance; both elements matter when lives and national continuity are at stake.

When the president crosses borders, protection expands into international collaboration. Navies and security detachments secure ports and hotels. Intelligence agencies exchange information. Local forces work with U.S. teams to secure routes and venues. Advance teams arrive days ahead, testing rooms, vetting staff, and ensuring that every handshake will be as safe as possible. Even simple things, such as the food served at state dinners, are vetted and tested. Nothing is left to chance.

Security is not only metal and code. It is also people who study patterns and prepare for the worst cases. Agents and officers work long hours, shifting between visibility and concealment. Their work is a quiet, relentless readiness. They balance the need for protection with the requirement that the president appear reachable and open. That paradox is not accidental. It is essential. The public must see a leader who can meet them and represent them, yet that meeting must be engineered to minimize risk.

This level of protection demands vast resources. Maintaining personnel, vehicles, aircraft, technology, and training takes money measured in the billions. Those costs are arguments about priorities and safety. They are also investments in continuity, in the idea that governance must survive even the most extreme shocks. There is always a tension between openness and safety, between civic ritual and defensive necessity. Democracies contend with that tension daily.

There is an irony at the heart of the role. The person whose office symbolizes liberty often lives within constraints that the public cannot fully imagine. Public appearances are rehearsed, movements are choreographed, and friendly crowds are observed for signs of danger. The president must project warmth and accessibility while existing inside a bubble of security. That tension shapes public life. It alters how decisions are made, how trust is built, and how the leader experiences the world.

Watching this system, I find myself thinking about the nature of modern power. The fortress around the president is built from invention: sensors, secure communications, armored vehicles, and trained teams. It is also built from a deeper human practice, the practice of imagining danger and working to stop it. Security is a form of applied thought, where imagination meets reality in the service of survival.

At the same time, we must ask what freedom we are willing to trade for safety, and how much secrecy is appropriate in a democracy that values openness. The answer changes with history, but the question remains. Protecting a leader is more than a technical challenge. It is a political and moral problem, and it reveals how societies balance trust and fear.

In the end, the man most guarded is also a mirror. The protective systems that surround him show us what we value, what we fear, and how we choose to preserve what matters. They are, at their best, a testament to human ingenuity and dedication, and at their most troubling, a reminder of the fragile edge on which public life stands.

Chapter 21
Accidents That Changed the World
When Mistakes Became Masterpieces

Human progress has often moved through intention, planning, and careful design. Yet some of our greatest breakthroughs appeared in moments where no breakthrough was expected at all. They arrived in a spill, a crack, a misplaced experiment, or a simple human oversight. These accidents remind us that invention is not only a matter of skill, but also of attention. They reveal how curiosity can turn missteps into turning points.

A Failed Glue That Held the World Together

In the early 1970s, Spencer Silver at 3M was trying to create a strong adhesive. What emerged instead was a glue that barely clung to anything. It was stable but weak, too light for industry and too odd for disposal. For years, it sat unused, a solution without a problem.

One of Silver's colleagues, Art Fry, sang in a church choir. Annoyed by bookmarks falling from his hymnal, he remembered the strange adhesive and applied it to pieces of scrap paper. The notes stuck, released, and stuck again. They held without tearing.

This experiment led to the Post-it Note, an invention that became a quiet tool of human memory. What began as a failed formula became a new way to capture passing thoughts, reminders, and ideas. It showed that value sometimes hides in gentleness rather than strength.

Radar Waves and an Unexpected Warm Meal

In 1945, Percy Spencer was testing radar equipment at Raytheon when a chocolate bar in his pocket softened. Instead of dismissing the heat as a nuisance, he questioned it. He placed popcorn kernels near the device. They burst. An egg exploded. The pattern was clear.

These observations led to the first microwave oven. A technology born from wartime research became a household appliance that transformed cooking. A wave once used to detect distant aircraft found its place warming dinners across the world. The shift from battlefield technology to everyday comfort began with a melted snack.

A Flicker of Light and a New Vision of the Body

In 1895, Wilhelm Conrad Roentgen was studying cathode rays when he noticed an unexplained glow on a fluorescent screen. Curious, he placed objects between the source and the screen and discovered that some materials allowed invisible rays to pass through them. When he held his own hand in front of the plate, the bones appeared as dark silhouettes.

This was the first X-ray. Within months, the technology spread across hospitals. Surgeons used it to locate fractures and bullets without opening the body. Medicine had gained a new way to see. The discovery emerged not from a planned breakthrough, but from Roentgen's insistence on investigating an unexpected light.

Nature's Burrs and a Fabric That Changes Shape

In 1941, Swiss engineer George de Mestral returned from a walk with his dog covered in tiny burrs. Instead of brushing them off, he studied them under a microscope. He saw hundreds of small hooks designed by nature to cling to fur.

Inspired, he engineered two cloth layers, one with loops and one with hooks. Pressed together, they held firmly; pulled apart, they separated easily. This became Velcro.

From a simple walk emerged a fastening technology used in shoes, clothing, and even space missions. A natural irritation turned into a global innovation.

A Falling Flask and the Birth of Safety Glass

In 1903, French chemist Édouard Benedictus accidentally dropped a glass flask. It cracked but did not shatter. The flask had been coated inside with cellulose nitrate, forming a thin, transparent layer that held the pieces together.

This accident inspired the development of laminated safety glass. It became essential for car windshields, airplanes, protective gear, and modern architecture. A near-mishap in a laboratory became protection for millions.

A Cylinder That Clogged and a Material That Slipped Free

In 1938, Roy Plunkett at DuPont was exploring refrigerant gases when he opened a cylinder that should have contained gas but instead held a white, waxy solid. The substance resisted heat, friction, and chemical reactions. It was polytetrafluoroethylene, later known as Teflon.

Its unique properties created a wide range of uses, from nonstick cookware to industrial seals and aerospace components. A blocked cylinder revealed one of the most versatile materials in modern engineering.

A Sweet Taste on Unwashed Hands

In 1879, chemist Constantin Fahlberg was investigating coal-tar compounds when he noticed that his dinner tasted unusually sweet. He traced the flavor to residue on his unwashed hands. That substance was saccharin, the first artificial sweetener.

This discovery reshaped food industries and medical treatments, offering new options for people with diabetes and those seeking low-calorie alternatives. A reminder to wash hands led to a new chapter in human diet.

A Spring That Refused to Sit Still

During World War II, naval engineer Richard James was working with springs designed to stabilize ship instruments. One of the springs fell from his desk and seemed to walk rather than collapse. The motion fascinated him.

His wife named the creation Slinky. It became a toy recognized around the world, proof that play can arise from unintended motion. A failed stabilizing device became a symbol of simple joy.

The Mold That Saved Millions

In 1928, Alexander Fleming returned from a holiday to find one of his Petri dishes contaminated with mold. Around the mold, bacteria had disappeared. He identified the mold as Penicillium notatum and the compound it produced as penicillin.

This accident became the foundation of modern antibiotics. It changed the outcome of infections and saved countless lives. A forgotten dish opened a new era in medicine.

When Party Tricks Became Compassion

In the early nineteenth century, people attended gatherings where they inhaled ether vapors for amusement. Observers noticed that participants seemed insensitive to pain while under its effects.

Physicians like Crawford Long and later William Morton recognized its potential for surgery. They refined the technique and demonstrated that operations could be performed without agony. The discovery transformed medicine. A recreational curiosity became a humane practice that continues to shape modern care.

A Quiet Pattern Across History

These stories reveal a pattern: progress is not always linear. It bends, hesitates, and sometimes arrives disguised as an inconvenience. Human curiosity converts accidents into advancements when someone chooses to ask why instead of turning away.

The world changes when mistakes are studied rather than ignored. In each case, a moment of surprise became an opportunity. These accidents were not simply errors. They were conversations between curiosity and chance. And in those conversations, new chapters of human knowledge were written.

Chapter 22
Jeff Bezos – The Relentless Architect of Tomorrow

Some people step through doors the world already opens. Others imagine an entirely new world and build the doorway themselves. Jeff Bezos belongs to the second kind. He did not merely ride the wave of the internet age. He helped shape its tide.

Bezos was born in 1964 in Albuquerque, New Mexico. From an early age, he showed a powerful curiosity for how things worked. He turned his family garage into a workshop of wires, tools, and unfinished experiments. He did not take things apart to break them, but to understand them. That search for how systems function would later define nearly everything he built.

At Princeton University, he studied electrical engineering and computer science. After graduating, he took jobs on Wall Street, where he learned how markets move and how data reveals hidden patterns. Yet even in the security of finance, he sensed that the future was shifting elsewhere. In 1994, he encountered a statistic that stopped him cold. Internet usage was growing at a staggering pace. To most people, it sounded like an abstract fact. To Bezos, it felt like a signal. He resigned from a lucrative career and drove from New York to Seattle, writing his business plan along the way. The vision was an online bookstore. The ambition behind it was far larger.

Amazon began in a garage with stacks of books and a simple website. Books were a strategic choice. There were too many titles for any physical store to carry, and an online shelf had no limits. From the beginning, Bezos focused relentlessly on the customer. Speed, choice, price, and trust became the pillars of the company. One by one, Amazon expanded into music, electronics, clothing, and nearly every product category imaginable. Shopping was no longer a destination. It became a quiet background act of daily life.

In 2007, Bezos introduced the Kindle. It was not just a device. It was a redefinition of reading. Thousands of books could live inside a slim screen. Words no longer waited on shelves. They traveled instantly. Publishing changed. Reading changed. The distance between author and reader shrank to a tap of the finger.

Another transformation happened almost silently. As Amazon grew, it needed enormous computing power to run its own operations. What began as internal infrastructure slowly revealed a wider opportunity. If Amazon needed scalable, flexible computing, so did thousands of other businesses. This insight gave birth to Amazon Web Services. Instead of purchasing expensive servers, companies could now rent computing power as easily as electricity. Startups scaled overnight. Entire industries migrated into the cloud. Today, much of the digital world rests on this invisible foundation.

Bezos also believed that the most natural way for humans to interact with machines was through speech. This belief led to the creation of Alexa and the Echo devices. Machines began to listen and respond. Homes grew quieter in an unusual way. Technology began to disappear into the background. Lights turned on with a word. Music arrived on command. Information surfaced without a screen. The oldest human interface, the voice, returned as the bridge between people and machines.

In 2000, Bezos extended his long view beyond Earth itself by founding Blue Origin. His goal was not a single launch or moment of spectacle. It was to make space travel sustainable, reusable, and open over time. The company followed a slow, methodical path, building rockets meant to land and fly again. For Bezos, space was not a race. It was a responsibility to the far future of humanity.

Across all these ventures runs a single thread: patience paired with relentless execution. Bezos did not build for applause. He built for scale. He accepted failure as part of the process. Experiments were expected to break. What mattered was learning faster than competitors and staying focused on the long horizon.

His influence reshaped not only how we shop, read, store information, and speak with machines, but also how companies think about growth. He showed that infrastructure can be as transformative as invention, and that the quiet systems beneath daily life often matter more than the devices in our hands.

Yet power on this scale also brings difficult questions. Amazon transformed labor, logistics, pricing, and competition across the globe. Its efficiency created convenience for millions while also raising concerns about market dominance, worker conditions, and the concentration of economic power. Blue Origin's vision of space travel invites both awe and debate about inequality and access to the future. Legacy is never single-edged.

When I reflect on Jeff Bezos, I do not see only an entrepreneur or a billionaire. I see a pattern of thought that defines our age. A belief that systems can always be rebuilt to move faster, reach farther, and serve more people. A willingness to bet on ideas long before proof arrives. And an understanding that the world does not change in a moment of inspiration, but through years of steady construction.

Bezos stands as a relentless architect of tomorrow. Not a dreamer who waits for the future to arrive, but a builder who assembles it piece by piece. His story reminds us that the shape of the modern world is not accidental. It is designed by those who are willing to imagine at scale, endure uncertainty, and build long after the excitement fades.

Chapter 23
The World's Largest Food Chain – How a Single Stand Became a Global Giant

The story begins small, with sunbaked asphalt and the smell of frying oil. In 1940, on a stretch of Route 66 in San Bernardino, California, Richard and Maurice McDonald ran a modest drive-in that served motorists passing through a town still recovering from the Depression. Their work was plain, practical, and direct: hamburgers, fries, milkshakes. Customers had grown used to waiting, to tray service delivered by carhops who balanced plates on car doors. The brothers watched those minutes pile up and saw an opportunity.

They asked a simple question: what if a meal could be ready the moment a customer asked for it? The answer was a sharper ritual of efficiency. They trimmed the menu to essentials, rearranged the kitchen like a small factory, and choreographed every motion. Buns were toasted, patties seared, and fries salted in a steady rhythm. Food moved toward the customer rather than the other way around. The change felt immediate, and it felt theatrical. Shiny stainless steel, open counters, and visible cooking gave customers sight, scent, and sound an invitation to witness the making of their meal.

The Speedee Service System

The McDonald brothers called their method the Speedee Service System. It was not merely faster; it was designed to be predictable. By standardizing processes, they reduced the chance of error. Every step had a place; every worker had a role. What had once been indulgence became a precise economy of motion and time. Their stand became a machine for delivering comfort, where convenience itself was the product.

The brothers did not set out to build an empire. They tuned their operation with the care of craftsmen, refining the timing of each burner and the placement of each spatula. Their aim was perfection at a single site rather than replication. Yet the rhythm they created attracted attention, and one visitor would hear a different music in that kitchen.

Ray Kroc, the multiplier

In 1954, a fifty-two-year-old salesman named Ray Kroc arrived at the San Bernardino stand in pursuit of a sale. He sold Multimixer milkshake machines and had made a life of convincing others to see value where they had not looked before. What caught his eye in that small kitchen was not the burgers, but the repeatable system. He saw potential for replication, for scale, for a promise that tasted the same in Des Plaines as it did in San Bernardino.

Kroc was not an inventor. He was a relentless organizer and a strategist of expansion. He insisted on strict standards, uniform training, and a contractual system that kept the brand consistent across locations. In 1955, he opened the first McDonald's franchise in Des Plaines, Illinois, and with that opening, he presented a new kind of comfort: wherever you sat, the hamburger would be familiar.

Kroc's life before McDonald's reads like a study in persistence. He had tried piano in smoky rooms, sold goods across the country, and endured businesses that failed. These experiences, however abrasive at the time, taught him how to persuade, how to endure, and how to recognize systems that could be multiplied. When he saw the McDonald brothers' kitchen, he heard a sequence of opportunities he knew how to amplify.

From local stand to global brand

Kroc bought the company outright in 1961 for $2.7 million, and the golden arches, designed by Stanley Meston, became a symbol of predictable comfort. As highways expanded and planes shrank distances, the red and yellow emblem traveled with the modern way of life. The menu adapted to local tastes without changing the underlying promise: speed, cleanliness, and sameness.

That promise proved powerful. Across continents, McDonald's learned to speak different culinary dialects: teriyaki in Japan, a potato-based patty in India, and croissants in parts of Europe, all while preserving the operational heartbeat laid down in that stand on Route 66. By offering the same basics everywhere, the company taught millions to trust a brand rather than a single cook.

Franchising became the engine of expansion, a legal and logistical architecture that turned a reliable method into an international template.

Cultural and economic ripple effects

Today, McDonald's serves tens of millions of customers every day in more than a hundred countries. Its reach is practical and symbolic. Practically, it reshaped supply chains, accelerated industrial food production, and made standardization a tool of efficiency. Symbolically, it became a mirror of globalization, showing how local habits can be connected to global systems.

The company's rise also forced hard questions. Standardization brought uniformity but sometimes eroded local food traditions. Franchising created business opportunities and also concentrated market power. The chain's global footprint raised debates about nutrition, labor practices, and environmental impact. The same architecture that brought predictability also demanded scrutiny and regulation.

A human moment in a machine

When I think about the McDonald brothers and Ray Kroc, I return to the sensory details: the hiss of a grill, the sweet steam of a milkshake, the small, precise movements of workers in a tight kitchen. Those images hold the truth about innovation. Change is often not a single heroic act but a tightening of routine until that routine becomes exceptional. The brothers asked a practical question and rearranged their world. Kroc heard the pattern and imagined it multiplied.

The story is also a reminder that inventions can outgrow their makers. The brothers perfected their stand and were content with craftsmanship; Kroc transformed that craft into scale. Each step had its costs and its gifts. The technique that delivered comfort to so many also reshaped how cities look, how food moves, and how people think about meals.

Why this matters

The rise of the world's largest food chain is not only a business case study. It is a lesson in systems thinking, in how a small improvement in process can scale into global influence.

It shows how culture and commerce entwine, and how a simple idea, sharpened by method and multiplied by vision, can alter daily life for millions.

When you see a McDonald's sign on a roadside or hear the familiar sizzle of a patty, you are witnessing the echo of a question asked in a modest kitchen: *What if* we made food move at the pace of life? That question set a revolution in motion whose consequences we still taste, in varied and complicated ways, today.

Thought That Fed the World

What began as one kitchen became a living metaphor for modern thought: the power of replication. A single efficient idea, refined, repeated, and scaled, fed the planet. It showed how human imagination doesn‘t always manifest in discovery; sometimes it lies in re-seeing the ordinary.

Two brothers reimagined the diner. One man reimagined the scale. Together, they changed not just how we eat, but how we think about possibility.

Every bite of a McDonald‘s burger carries the echo of that first what if‖ whispered in a small California stand, a thought that turned hunger into innovation and routine into ritual.

Chapter 24
The Business of Trading Emotions

We once traded goods. Then we traded gold. Today, we trade something far more delicate and deeply human: emotions. From the glow of ancient campfires to the glowing screens we cradle in our hands, we have always exchanged more than objects. We have exchanged awe and comfort, fear and excitement, belonging and belief. The storyteller in a small village once earned a meal by moving hearts. An actor on a wooden stage drew power from the applause that poured like waves from the crowd. Advertisers in the industrial age learned a subtle art. They stopped selling soap and started selling confidence. They sold not a product, but the feeling of being seen.

Now, the merchant of emotion has no heartbeat. It has no voice. It is an algorithm that listens more closely than any friend and studies us more deeply than we study ourselves. Platforms that promised connection have become vast emotional markets, filled not with goods but with reactions. A click becomes desire. A like becomes validation. A comment becomes a spark that fuels the engine of attention. Joy is measured in tiny glowing hearts. Outrage climbs to the top of every page because anger spreads faster than calm. What we feel is no longer simply human. It has become a commodity.

The trade is silent, but its scale is immense. The more intensely we feel, the more valuable we become, not as whole people but as patterns in a database.

We are profiles feeding a system that predicts what we will want before we want it. And so a question rises quietly within us. Will artificial intelligence expand our ability to feel, or will it reduce emotion to something predictable and programmable? Can empathy be awakened by a machine, or will it be replaced by simulation?

In a world where every feeling is captured and calculated, the bravest act may be to reclaim our inner world. To feel without being measured. To express without being tracked. Because even in a global economy of emotion, there is still one currency no code can counterfeit: a genuine human heart.

Chapter 25
The Quiet Revolution of Access: How the World Became Democratic Without Elections

Some revolutions announce themselves with drums and banners. They overturn thrones and redraw maps. Others unfold more quietly, reshaping the rules of who can speak, learn, and create. This is one of those quieter upheavals.

Over the past century, the world has experienced a steady, far-reaching change: the democratization of knowledge, creativity, memory, sound, and sight. It did not arrive as a single moment. It arrived in many small, persistent movements: a new printing technology here, a storage medium there, a global network that connected them all. Pixel by pixel, byte by byte, the modern public square expanded until anyone with curiosity and a device could join.

Knowledge: from guarded rooms to open pages

For most of human history, learning required proximity: to a teacher, to a city, to a library. Clay tablets and temple records, scrolls in Alexandria and Nalanda, private collections, and elite classrooms kept information on the inside and kept many on the outside. Knowledge was not only power; in practical terms, it was property.

Then, technologies changed the geometry of access. The printing press made mass literacy possible. Broadcast media extended a few voices to many. The internet multiplied those voices into billions. Search engines taught us how to ask, and online courses carried universities into homes far from campus. Today, AI tools can help a student draft an essay or a retiree explore a new subject with step-by-step guidance.

This is not to say the playing field is fully level. Digital divides persist. Language barriers remain. But compared with a century ago, a farmer with a smartphone can now reach the global library and the same basic resources once available only to a privileged few. That shift matters morally and practically: learning has started to feel like a common inheritance, not a scarce advantage.

Music: from palace salons to pocket players

Music used to travel through narrow halls. Composers performed for patrons, and audiences relied on gatekeepers to hear new sounds. Each leap in technology, from gramophones, radio, cassette, CD, and streaming, lowered the cost of listening and performing. Today, a phone contains more recorded music than could fit in any royal archive.

But there is a subtler change than quantity. The tools of production have also become accessible. Affordable recording gear, intuitive software, and algorithmic assistance let people craft songs that would have once required expensive studios. A busker with a melody can reach millions. The cultural consequence is that taste is no longer only curated by institutions; it emerges from communities.

Photography and video: everyone is a witness

Early photography demanded chemistry, time, and money. Capturing an image was a technical privilege. Kodak's promise that anyone could press the button began the redefinition. Digital sensors, then smartphones, put photography and filmmaking in the hands of billions. Platforms that distribute video made a global audience possible for a home-recorded clip.

The result is not merely more images. It is a richer visual record of human life by the people who live it. Personal archives, citizen journalism, and grassroots storytelling now coexist with professionally produced work. That coexistence changes what we remember and how we understand events, often in urgent and uncomfortable ways.

YouTube and the redistribution of attention

If the printing press made text portable, platforms like YouTube made vision portable. Television channels once decided who would be seen. Now, a person with a phone can teach, perform, report, and persuade. This change redistributed attention, not just content. It allowed niche expertise and unconventional voices to find audiences without permission from traditional gatekeepers.

The consequence is double-edged. Accessibility has amplified learning, creativity, and connection.

It has also multiplied misinformation and reduced the authority of traditional filters. The social task now is to build civic habits and technical systems that favor truth and accountability while preserving openness.

Writing and publishing: the end of the single gatekeeper

Publishing once required the assent of editors and houses that decided which voices reached readers. Self-publishing, blogs, and digital distribution changed that dynamic. Writers can now publish directly, reach readers globally, and iterate quickly. That liberation reshapes literary culture and invites new forms of exchange between author and audience.

Other realms transformed

Education. Massive open online courses and recorded lectures make higher learning available beyond campus walls. A motivated learner can pursue rigorous study at a fraction of the historical cost.

Entrepreneurship. A single smartphone can become the essential infrastructure for a small business, lowering the barrier to entry and enabling micro-entrepreneurs around the world.

Art and design. Tools that once required years of apprenticeship now provide immediate creative possibilities, expanding who can express and produce.

Communication and innovation. Instant networks accelerate collaboration. Open-source projects show how many minds, working without central command, can build complex systems such as Linux, Python, and global knowledge repositories like Wikipedia.

What democratization truly means

This revolution is more than technical diffusion. It is a philosophical shift. It removes or loosens the old gatekeepers: institutions that once decided who could learn, who could publish, who could be heard. It changes the moral baseline. Creativity is no longer a privilege of the few; it is a possibility for many.

That said, inclusion is incomplete. Algorithms shape visibility. Commercial platforms mediate access through business models. And the capacity to use these tools effectively still depends on education and infrastructure. Democratization is an ongoing process, not a finished victory.

Why this quiet revolution matters

The transformation reshapes daily life. It alters how people learn, work, be creative, and remember. It spreads the means of cultural production outward from centers to peripheries. It permits new forms of solidarity and new forms of conflict. It gives more people the tools to shape their future, but it also asks more of societies: how to govern shared spaces, protect privacy, and cultivate public-minded information habits.

A final note from the author

When I think of this change, I see small acts that become movements. A child in a remote village discovering a science video online. A local musician uploading a song that becomes a friend's favorite in another country. A community documenting its own history rather than waiting for someone else to tell it. Those moments add up into a different world, quieter than a revolution yet no less consequential.

This quiet democratization asks us to be intentional. If access is now widespread, how will we teach judgment? If anyone can speak, how will we cultivate listening? The revolution has given us tools. The next task is to use them well.

Chapter 26
Dark Factories: When We Switched Off the Lights

The world has always been illuminated by human labor. Fire once glowed in the heart of blacksmith workshops. Lanterns followed weavers' hands through long nights. Later, the modern age flooded factories with bright fluorescent lights that hummed above the motion of machines and the effort of millions. Light became the quiet companion of work. It revealed what hands created and what minds imagined. It watched as humans transformed raw materials into the shape of progress.

Then came a radical question. *What if* a factory no longer needed light at all

Not because night conquered day. Not because electricity failed. But because there were finally no human eyes inside the building. The workers stepped away, and their tools stayed behind to continue the task. Machines became the sight. Robots became the hands. Algorithms set the rhythm. These places did not become abandoned ruins. They became something new. They became dark factories.

What exactly is a dark factory?

A dark factory, often called lights-out manufacturing, is a place where production carries on without a single person walking the floor. Every activity, from assembling components to inspecting quality and packaging the finished product, is guided by automation. With no human eye needing brightness and no human body needing comfort, the lights remain off, and the air remains unconditioned. Productivity does not pause for day or night. The clock loses authority. Work continues as long as code tells machines to move.

Inside these spaces, the factory no longer follows the rhythm of breathing workers but the clean precision of engineering. Sensors watch what humans might miss. Metal arms repeat movements with unwavering patience. Artificial intelligence studies patterns and responds faster than any manager could. Industrial imagination reaches its purest form.

Origins: from dream to demonstration

The earliest vision of a factory without people lived in the minds and fiction. After the Second World War, new machines and new ideas led dreamers to wonder how far automation could go. By the 1980s, automobile companies in Detroit tried to move toward factories filled with robots. Yet the tools of the time could not master unexpected changes. Human judgment was still needed to handle imperfection.

The breakthrough appeared quietly in Japan. At the foot of Mount Fuji, the robotics company FANUC constructed a plant capable of running with no one inside. Robots assembled other robots in continuous operation that lasted for weeks. The factory was silent, dark, and astonishing. This was no longer an imaginative prediction. It was a demonstration of what human thought could build.

Not long after, Europe pushed the idea further. Philips opened an automated plant in the Netherlands where robots crafted electric shavers with remarkable accuracy. The success of such sites inspired leaders to believe that manufacturing no longer had to depend on cheap global labor. Robotics could bring production home again.

Who leads today

The new measure of industrial power is robot density, meaning how many robots work in a country for every ten thousand human workers. South Korea holds the highest position in the world, with Singapore close beside it. Japan and Germany remain strong leaders in automation. China, once famous for low-cost human labor, has now charged forward into a future of automated labor. Its investments in robotics and artificial intelligence aim to reshape it into a global center of autonomous manufacturing.

The evolution: a timeline of thought

1950s to 1980s: Science fiction and experimental robotics planted the first seeds.

1990s: Automated islands emerge with limited independence for night shifts.

2000s: Integration of systems allows full lines to operate together.

2010s: Digital control and smart robotics compete with overseas labor.

2020s: Nations scale automation to support energy storage, electronics, and electric vehicles that must be built every hour of the day.

Impact on society

Dark factories do not remove the need for work. They change what work looks like. Human hands step back from repetitive tasks, but new responsibilities rise for robot technicians, programmers, and data analysts. For some, this shift feels unsettling. Long valued industrial jobs fade and identity with them. For others, the change feels freeing.

Labor that once demanded physical strain and constant danger becomes safer and smarter.

Yet within this transformation, a question remains. What happens to purpose when machines perform the motions that once filled our days? What does progress mean when productivity no longer depends on the presence of people? There is a balance to be found between efficiency and humanity. Thought must guide the tools we build.

Envisioning the future: when the world sleeps, factories dream

Imagine standing in a city at midnight. The streets are quiet. Windows are dark. Silence spreads like soft mist. But inside automated plants, creation has not paused. Components slide into position. Sparks flash in hidden corners. Electric motors hum through the early hours. The world rests while industry keeps breathing.

The remarkable part is not the darkness. It is the human imagination that created it. Someone dared to turn the lights off. Someone believed work could continue without a witness. A thought sparked it, and technology followed.

Every dark factory is more than a structure of steel. It is a symbol of how far the human mind is willing to travel. We picture a new possibility and then build it. If robots can carry the tools of creation, then humans are challenged to redefine the meaning of contribution.

Innovation has always been a quiet wave moving forward. A single idea grows until it reshapes the familiar. And when we step back, the world looks different than before.

Dark factories invite us to think deeply about the next chapter of progress. The lights may be off inside them, but the light of human thought shines brighter than ever.

Conclusion: The Eternal Wave

Every wave begins as a ripple, a tremor so faint it might pass unnoticed, yet carrying the seed of a tide, the whisper of a storm, the quiet strength to move an ocean.

Human thought has always begun this way. The spark that lit the first fire, the wheel rolling across ancient dust, the press of ink upon paper, the hum of a hidden circuit, all were born as fragile murmurs of possibility. Once released, they swelled into tides that carried civilizations forward, reshaping how humanity dreamed, built, and became.

Look back across the centuries, and you won‘t see isolated flashes of genius, but an unbroken current of wonder, one mind igniting another, one question answering the next. The Wright brothers were never just bicycle makers with a bold idea; they were heirs to da Vinci‘s sketches, to Chinese kites tugging against the wind, to the mathematicians who mapped the invisible lift of air. Kitty Hawk was not an accident of time; it was the crest of a wave gathering strength for centuries.

The same pulse runs through the digital age. The microchip did not appear from nowhere; it rose from Leibniz‘s binary dream, from Ada Lovelace‘s elegant logic, from Turing‘s wartime code and the sleepless engineers who soldered imagination into silicon. Every thought became current, feeding the next an invisible network of minds stretching across generations.

Thoughts are never still. They are rivers winding, merging, breaking apart, and rejoining again. They evaporate into dreams, condense into ideas, and fall once more as inventions. Nothing is ever lost, only transformed. This is humanity's true rhythm: imagination flowing forward like water, restless, alive, eternal.

Empires rise, crumble, and return to dust; monuments fade under wind and time. Yet the smallest idea outlives them all. Mesopotamian palaces lie in ruin, but their symbols still speak. Rome's legions vanished, but its arches, roads, and laws still shape our cities. What survives of us will never be the gold we hoarded, but the questions we dared to ask and the sparks we passed on to others.

And even now, waves are forming. Somewhere, a child gazes at the night sky and wonders how to touch the stars. A scientist peers into a microscope, asking if consciousness can be born in a circuit. An artist sketches a form that may someday guide a robot's hand. Every quiet curiosity is already a ripple, and history reminds us that ripples never stay small.

The lesson is timeless and urgent: we are not bystanders on a shore; we are part of the tide. Every curiosity you nurture, every question you refuse to silence, every dream you dare to follow becomes a current in this vast sea of thought. No ripple is meaningless.

Who could have guessed that a toy helicopter in two boys' hands would teach us to fly? That scribbles in a notebook would bend the fabric of the universe? That an apple falling from a tree would redefine gravity itself?

Greatness is not the birthright of the few. It is born wherever courage meets imagination in classrooms, in workshops, in quiet minds that refuse to settle for what is known. Edison was right: genius is ninety-nine percent effort. But that last one percent spark is where all progress begins. It is thought that refusing to be still, curiosity refusing to fade, the soul insists that there is always more to see.

So let this be the echo you carry: humanity‘s story is unfinished, and it must remain so. We are not defined by what we own, but by what we imagine. From firesides to laboratories, from myths whispered in caves to knowledge etched in quantum code, we have ridden the waves of thought across time‘s boundless ocean, and the tide is still rising.

The responsibility is ours: to keep asking, to keep daring, to let our thoughts ripple outward for those who will come after. One day, they will look back and ask what waves we set in motion. May the answer be worthy of the questions that began it all.

For in the end, no empire endures, no monument escapes erosion, no fortune survives time. What lasts, what truly endures are the waves of thought: endlessly carried, endlessly reborn, whispering through minds not yet born.

Think, and the world will change.

Epiphanies from This Wave

PART 5 – Thoughts That Built Civilization

— On an Italian bench, moneylenders exchanged trust. When the bench broke, trust shattered, too, and "bankruptcy" entered our tongue.

— Goats leapt with wild energy after chewing red berries. From their frenzy came coffee, and from coffee came revolutions of thought in cafés.

— A simple tea leaf steeped in water toppled thrones, fuelled wars, and bound continents together in trade.

— A bead of glass once worn as jewellery, learned to carry light and, in time, became the veins through which the digital world breathes.

— When ice was trapped and tamed, food survived beyond seasons. Refrigeration turned survival into abundance, and civilization breathed easier.

Epilogue

There was a time when we owned nothing but wonder.

No tools, no language, no maps to tell us where we stood beneath the stars. Only bare feet upon the earth and a mind restless with questions. Survival was our first thought: keep warm, find food, endure the night. Yet somewhere in the silence of an ancient cave, a spark leapt from imagination to flame. A human hand reached toward darkness and whispered, *What if*? Fire was not granted by the gods; it was coaxed from curiosity. And in that instant, thought became a tool and the tool, destiny.

From that ember, imagination began its endless climb. We painted the walls with our hopes and hunts, leaving shadows that spoke when words did not yet exist. Clay learned to carry meaning; lines became language, and memory found a home. We sowed seeds and waited for an act of faith that the unseen could grow. Villages rose, then cities, then civilizations, each built upon the quiet daring of another mind that asked why not.

We looked up. The stars no longer frightened us; they guided us. Constellations became calendars. The Greeks turned wonder into reason, chiseling questions into philosophy. The scholars of the Islamic dawn stitched wisdom across continents, teaching light to bend, numbers to sing, and medicine to heal. Gutenberg gave thought to the wings of paper. Galileo tilted the heavens. Newton taught motion its law, and steam carried the dream forward into the age of iron and fire.

Wires crept beneath oceans, and the world began to speak to itself. A pulse crossed continents: Can I send my voice across distance? The telegraph answered yes. Then came Einstein bending time, Curie distilling the invisible, Turing giving reason to electricity, and the DNA helix revealing the code that writes us all. Rockets rose, and thought left the cradle of Earth.

Now, our machines hum with borrowed imagination. They paint, compose, and write not as creators, but as reflections of the same human wonder that first lit a cave. Ahead shimmer questions that refuse silence: Can we heal age? Upload a soul? Defy the end?

We don‘t yet know. But knowing was never the point. Our greatness lies in the courage to ask.

For the lightning that shaped our world never fell from the sky, it has always lived inside the human mind.

And as long as thought flickers, the waves will rise.

Acknowledgements

To my son — your steady encouragement has been a quiet flame through every long night of writing. Your belief in me gave strength to every page that might never have been written without it.

To my family and friends, thank you for standing beside me with patience and honesty. You read the early drafts, pointed out what I could not see, and reminded me why this journey mattered. Each suggestion, each word of faith, helped this book find its clarity and heart.

To those whose unseen efforts shaped these final pages — the designers, formatters, editors, and collaborators who brought structure and elegance to thought — your craft turned imagination into something tangible. You helped transform reflection into art.

I extend my deepest respect and gratitude to Dr. Duvvuri Subbarao for his gracious foreword. His careful reading, nuanced reflections, and distinguished endorsement have brought an honour to this work that I could never have assumed. His voice enriches these pages and strengthens the purpose with which they are given to the world.

I also extend gratitude to the quiet intelligence of the digital age — the tools that assisted in research, refinement, and reflection. They did not replace human thought; they reminded me that technology itself is born from it. Every keystroke was guided by the same curiosity that once struck flint against stone.

Finally, to everyone who believes that ideas matter, that questions have weight, and that imagination deserves time — thank you. It is for minds like yours that this book was written, and it is in such minds that the next wave of thought will arise.

Annexure 1

When Zero Became Everything — The Rise of Eight World-Changing Inventions

Invention	Inventor / Origin	Year (Approx.)	Years Ago (from 2025)	What the World Started With	What the World Has Today
Printing Press	Johannes Gutenberg	~1440	~585 years ago	Zero printed books	Over **1 trillion** books printed since Gutenberg
Steam Engine	James Watt (improved engine)	1776	~249 years ago	Zero locomotives; zero powered factories	Millions of engines; foundation of the Industrial Revolution
Photography	Louis Daguerre	1839	~186 years ago	Zero photographs	**~1.6 trillion** photos captured every year
Telephone	Alexander Graham Bell	1876	~149 years ago	Zero phone calls	**~6.5 billion** active connections
Electric Light	Thomas Edison	1879	~146 years ago	Zero lightbulbs	Over **25 billion** lightbulbs glowing
Automobile	Karl Benz	1885	~140 years ago	Zero cars on Earth	**~1.45 billion** cars in use
Radio	Guglielmo Marconi	1895–1901	~125–130 years ago	Zero radio receivers	Over **3 billion** radio devices
Airplane	Wright Brothers	1903	~122 years ago	Zero seconds of human flight	**~100,000 flights daily; 900,000+ aircraft built**

References

Annexure: When Zero Became Everything — The Rise of Eight World-Changing Inventions

Printing Press

1. Britannica. (2024). *Johannes Gutenberg and the invention of the printing press.* Encyclopaedia Britannica.
2. Eisenstein, E. (1980). *The printing press as an agent of change.* Cambridge University Press.
3. UNESCO. (2021). *Global book production statistics.* UNESCO Publishing Report.

Steam Engine

1. Britannica. (2024). *James Watt and the development of the modern steam engine.* Encyclopaedia Britannica.
2. Landes, D. (1969). *The unbound Prometheus: Technological change and industrial development in Western Europe.* Cambridge University Press.
3. *Photography*
4. Newhall, B. (1982). *The history of photography.* The Museum of Modern Art.
5. Smithsonian Institution. (2023). *Daguerre and the birth of photography.* National Museum of American History.

Telephone

1. International Telecommunication Union. (2023). *Global communication statistics: Mobile and fixed-line connections.* ITU Publications.
2. Library of Congress. (2024). *Alexander Graham Bell and the invention of the telephone.* U.S. Government Archives.

Electric Light

1. Edison Papers Project. (2024). *Thomas Edison's electric light innovations.* Rutgers University Press.
2. IEEE Global History Network. (2023). *Development of electric lighting technology.* Institute of Electrical and Electronics Engineers.

Automobile

1. Britannica. (2024). *Karl Benz and the invention of the automobile.* Encyclopaedia Britannica.
2. International Organization of Motor Vehicle Manufacturers. (2023). *World vehicle in-use statistics.* OICA Annual Report.

Radio

1. IEEE Spectrum. (2022). *Marconi and the origins of radio communication*. IEEE.
2. UNESCO. (2022). *World radio statistics and global broadcast reach*. UNESCO Media Report.

Airplane

1. FlightRadar24. (2023). *Daily global commercial flight volume*. FlightRadar Analytics.
2. International Civil Aviation Organization. (2023). *Annual global aviation statistics*. ICAO Statistical Yearbook.
3. Smithsonian National Air and Space Museum. (2024). *Wright Brothers: The first powered flight*. Smithsonian Press.

Introduction

1. Deacon, T. W. (1997). The symbolic species: The co-evolution of language and the brain. W. W. Norton & Company.
2. Diamond, J. (1997). Guns, germs, and steel: The fates of human societies. W.W. Norton & Company.
3. Fagan, B. M. (2004). People of the Earth: An Introduction to World Prehistory (12th ed.). Prentice Hall.
4. Harari, Y. N. (2014). Sapiens: A brief history of humankind. Harper.
5. Mithen, S. (1996). The prehistory of the mind: The cognitive origins of art, religion, and science. Thames & Hudson.
6. Tattersall, I. (2012). Masters of the planet: The search for our human origins. Macmillan.
7. Wynn, T., & Coolidge, F. (2009). How to think like a Neandertal. Oxford University Press.

Part 1

Chapter One

1. Damasio, A. (1994). Descartes' Error: Emotion, Reason, and the Human Brain. New York:
2. G.P. Putnam's Sons.
3. Dawkins, R. (1976). The Selfish Gene. Oxford: Oxford University Press.
4. Gazzaniga, M. S., Ivry, R. B., & Mangun, G. R. (2019). Cognitive Neuroscience: The Biology of the Mind (5th ed.). New York: W. W. Norton & Company.
5. LeDoux, J. (2002). Synaptic Self: How Our Brains Become Who We Are. New York: Viking.
6. Libet, B. (1985). Unconscious cerebral initiative and the role of conscious will in voluntary action. Behavioral and Brain Sciences, 8(4), 529–566.
7. When a Spark Becomes a Symphony

Chapter Two

1. Crystal, D. (2010). The Cambridge Encyclopedia of Language. Cambridge University Press.
2. Deutscher, G. (2010). Through the Language Glass: Why the World Looks Different in Other Languages. Metropolitan Books.
3. Ethnologue. (2024). Languages of the World: Summary by language family. SIL International. Retrieved from
4. Lieberman, P. (2007). The Evolution of Human Speech: Its Anatomical and Neural Bases. Current Anthropology, 48(1), 39–66.
5. McWhorter, J. (2011). The Power of Babel: A Natural History of Language. Harper Perennial.
6. Pinker, S. (1994). The Language Instinct: How the Mind Creates Language. William Morrow and Company.
7. The First Sunrise: Language, the Bridge of Thought

Chapter Three

1. Eisenstein, E.L. (1980). The Printing Revolution in Early Modern Europe. Cambridge University Press.
2. Evolution of Storytelling
3. Lanham, R.A. (2006). The Electronic Word: Democracy, Technology, and the Arts. University of Chicago Press.
4. McLuhan, M. (1962). The Gutenberg Galaxy: The Making of Typographic Man. University of Toronto Press.
5. Standage, T. (2007). The Victorian Internet: The Remarkable Story of the Telegraph and the Nineteenth Century's On-Line Pioneers. Walker & Company.
6. The Epic of Gilgamesh. (n.d.). Translated by N.K. Sanders. Penguin Classics, 2002.

Chapter Four

1. Evolution of Art and Painting: The Thought of Visually Expressing Emotions. Gombrich, E. H. (1995). The Story of Art (16th ed.). London: Phaidon Press. Honour, H., & Fleming, J. (2009). A World History of Art (7th ed.). London: Laurence King Publishing.
2. Kleiner, F. S. (2016). Gardner's Art through the Ages: A Global History (15th ed.). Boston, MA: Cengage Learning.
3. Lucie-Smith, E. (2003). Art Today. London: Phaidon Press.
4. Sayre, H. M. (2012). A World of Art (7th ed.). Boston, MA: Pearson Education. Gombrich, E. H. (1995). The Story of Art (16th ed.). Phaidon Press.
5. Honour, H., & Fleming, J. (2009). A World History of Art (7th ed.). Laurence King Publishing.

6. Kleiner, F. S. (2016). Gardner's Art through the Ages: A Global History (15th ed.). Cengage Learning.
7. Lucie-Smith, E. (2003). Art Today. Phaidon Press.
8. Sayre, H. M. (2012). A World of Art (7th ed.). Pearson Education.

Chapter Five

1. Boethius, A. M. (1989). Fundamentals of Music: Anicius Manlius Severinus Boethius. Yale University Press.
2. Copland, A. (1952). What to Listen for in Music. McGraw-Hill.
3. Grout, D. J., & Palisca, C. V. (2006). A History of Western Music (7th ed.). W.
4. W. Norton & Company.
5. Howard, P. (2002). The Birth of the Orchestra: History of an Institution, 1650– 1815. Oxford University Press.
6. Isaacson, W. (2011). Steve Jobs. Simon & Schuster.
7. Music and Its Evolution
8. Sachs, C. (1943). The Rise of Music in the Ancient World. W. W. Norton & Company.
9. Taruskin, R. (2005). The Oxford History of Western Music. Oxford University Press.
10. Temperley, N. (Ed.). (1983). The Romantic Age: 1800–1914. Macmillan.
11. Watson, I. (2010). Song and Symbol: Ancient Music and Spiritual Meaning. University of California Press.

Chapter Six

1. Bale, J., & Christensen, M. K. (2004). Post-Olympism? Questioning Sport in the Twenty-First Century. Berg Publishers.
2. Boyle, R., & Haynes, R. (2009). Power Play: Sport, the Media and Popular Culture. Edinburgh University Press.
3. From Sticks and Stones to Stadiums
4. Guttmann, A. (2004). From Ritual to Record: The Nature of Modern Sports. Columbia University Press.
5. Hutchins, B., & Rowe, D. (2012). Sport Beyond Television: The Internet, Digital Media and the Rise of Networked Media Sport. Routledge.
6. Mandell, R. (1976). The First Modern Olympics. University of California Press. Wenner, L. A. (Ed.). (1998). MediaSport. Routledge.
7. Young, D. C. (2004). A Brief History of the Olympic Games. Wiley-Blackwell.

Chapter Seven

1. Purves, D., Augustine, G. J., & Fitzpatrick, D. (2001). Neuroscience (2nd ed.). Sunderland, MA: Sinauer Associates.
2. Rossing, T. D. (2007). The Science of Sound (3rd ed.). Pearson Education. Goldstein, E. B. (2013). Sensation and Perception (9th ed.). Cengage Learning.

3. Greene, B. (1999). The Elegant Universe: Superstrings, Hidden Dimensions, and the Quest for the Ultimate Theory. W. W. Norton & Company.
4. Levitin, D. J. (2006). This Is Your Brain on Music: The Science of a Human Obsession. Dutton.
5. NASA – Sound and Vibration Basics. Retrieved from https://www.nasa.gov. The Symphony of Vibrations – How We Listen to the World.

Chapter Eight

1. Kinoshita, S., Yoshioka, S., & Miyazaki, J. (2008). Physics of structural colours. Reports on Progress in Physics, 71(7), 076401.
2. Prum, R. O., & Torres, R. H. (2003). Structural colour production by quasi-ordered nanostructures in avian feather barbs. Journal of Experimental Biology, 206(14), 2409–2429.
3. Potyrailo, R. A., et al. (2007). Morpho butterfly wing scales demonstrate a highly selective vapour response. Nature Photonics, 1(2), 123–128.
4. Parker, A. R., & Townley, H. E. (2007). Biomimetics of photonic nanostructures. Nature Nanotechnology, 2(6), 347–353.
5. Vukusic, P., & Sambles, J. R. (2003). Photonic structures in biology. Nature, 424(6950), 852–855.
6. The Peacock's Feather – Where Light Learns to Dance

Chapter Nine

1. Bartlett, N. W., et al. (2015). A 3D-printed, functionally graded soft robot powered by combustion. Science, 349(6244), 161–165.
2. Cianchetti, M., Calisti, M., Margheri, L., Kuba, M., & Laschi, C. (2015). Bioinspired locomotion and grasping in water: The soft eight-arm OCTOPUS robot. Bioinspiration & Biomimetics, 10(3), 035003.
3. Laschi, C., Mazzolai, B., & Cianchetti, M. (2016). Soft robotics: Technologies and systems pushing the boundaries of robot abilities. Science Robotics, 1(1), eaah3690.
4. Marchese, A. D., Katzschmann, R. K., & Rus, D. (2015). A recipe for soft fluidic elastomer robots. Soft Robotics, 2(1), 7–25.
5. The Creature Without Bones That Taught Us to Build Robots
6. Whitesides, G. M. (2018). Soft robotics. Angewandte Chemie International Edition, 57(16), 4258–4273.

Chapter Ten

1. Bascom, W. (1964). Waves and Beaches. Cousteau, J. Y. (1953). The Silent World.
2. Fernández-Armesto, F. (2003). Pathfinders: A Global History of Exploration. Ocean: The Mysterious Blue That Drew Us In

Chapter Eleven

1. Bolt Threads – https://boltthreads.com
2. Lewis, R. V. (2006). —Spider silk: ancient ideas for new biomaterials. Chemical Reviews, 106(9), 3762–3774.
3. NASA Orb Weaver Experiment – Nexia Biotechnologies (2002). Biosteel™ project archives.
4. Rising, A., & Johansson, J. (2015). —Toward spinning artificial spider silk. Nature Chemical Biology, 11(5), 309–315.
5. Science News (2012). —Could a Spider Web Really Lift the Statue of Liberty? – https://www.sciencenews.org
6. THE THREAD OF GENIUS
7. When Humans Looked at the Spider and Saw the Future Tufts University Silk Lab – https://silklab.tufts.edu/
8. Vepari, C., & Kaplan, D. L. (2007). —Silk as a biomaterial. Progress in Polymer Science, 32(8–9), 991–1007.
9. Victoria & Albert Museum – Golden Spider Silk Exhibition (2012)
10. Vollrath, F., & Knight, D. P. (2001). —Liquid crystalline spinning of spider silk. Nature, 410(6828), 541–548.

Chapter Twelve

1. Barber, E. J. W. (1994). Women's Work: The First 20,000 Years - Women, Cloth, and Society in Early Times. W. W. Norton & Company.
2. Isaacson, W. (2011). The Innovators. Simon & Schuster.
3. Jenkins, D. (2003). The Cambridge History of Western Textiles. Cambridge University Press.
4. National Geographic Society. (2020). —History of Textiles. Retrieved from nationalgeographic.com
5. References and Bibliography
6. Schoeser, M. (2007). World Textiles: A Concise History. Thames & Hudson.
7. Smithsonian Institution. (2023). —Looms and Innovation. Smithsonian Archives.
8. When Humans Decided to Clothe Themselves
9. Yafa, S. (2005). Big Cotton: The Biography of a Revolutionary Fiber. Penguin Books.

Chapter Thirteen

1. Canfora, L. (1986). The Vanished Library: A Wonder of the Ancient World. University of California Press.
2. Frame, G. (2007). The Archives of Assyrian Kings, 8th–7th Centuries BCE. Neo-Assyrian Text Corpus Project.
3. Hunter, D. (1943). Papermaking: The History and Technique of an Ancient Craft. Knopf.

4. Kramer, S.N. (1963). The Sumerians: Their History, Culture, and Character. University of Chicago Press.
5. Marshack, A. (1972). The Roots of Civilization. McGraw-Hill. The Thought That Froze Time: From Clay to Paper
6. Tsien, T.-H. (1985). Paper and Printing. In Science and Civilisation in China, Volume 5, Part 1. Cambridge University Press.

Chapter Fourteen

1. Barrett, T. H. (2009). The Woman Who Discovered Printing. Yale University Press.
2. Needham, J. (1986). Science and Civilisation in China, Vol. 5: Chemistry and Chemical Technology, Part 1: Paper and Printing. Cambridge University Press.
3. The Man Who Gave Thought a Medium
4. Tsien, T.-H. (1985). Paper and Printing, in Science and Civilisation in China (Vol. 5, Part 1). Cambridge University Press.

Chapter Fifteen

1. Darnton, R. (2009). The Case for Books: Past, Present, and Future. PublicAffairs.
2. Eisenstein, E. L. (1979). The Printing Press as an Agent of Change. Cambridge University Press.
3. Isaacson, W. (2011). Steve Jobs. Simon & Schuster.
4. Lebert, M. (2009). A Short History of eBooks. Retrieved from University of Toronto eBook Library.
5. Man, J. (2002). Gutenberg: How One Man Remade the World with Words. Wiley.
6. Pettegree, A. (2014). The Book in the Renaissance. Yale University Press. Printing — When Thought Multiplied

Chapter Sixteen

1. Alhazen (Ibn al-Haytham), Kitab al-Manazir (Book of Optics), 11th century. Eyewear – The Thought That Helped Humanity See Better
2. Goss, D. A. (1998). A Concise History of Optometry. Southern College of Optometry.
3. Hockenberry, J. (2013). —The Spectacular History of Eyeglasses, Wired.
4. Ilardi, V. (2007). Renaissance Vision from Spectacles to Telescopes. American Philosophical Society.
5. Isaacson, W. (2011). Benjamin Franklin: An American Life. Simon & Schuster.
6. Watson, A. (2014). The Evolution of Eyeglasses. Journal of the American Optometric Association.

Chapter Seventeen

1. Galileo, G. (1610). Sidereus Nuncius [Starry Messenger].
2. Hubble, E. (1936). The Realm of the Nebulae. Yale University Press.

3. National Aeronautics and Space Administration (NASA). (1990). The Hubble Space Telescope: A Journey Through Time.
4. References and Bibliography
5. The Instruments of Discovery: Telescopes and Microscopes
6. van Leeuwenhoek, A. (1677). Letter to the Royal Society of London, detailing his discoveries of —animalcules‖.
7. Webb, J. (2021). The James Webb Space Telescope: A New Window into the Universe. NASA.

Chapter Eighteen

1. Eastman, G. (1926). Memories of the Birth of Kodak.
2. Gernsheim, H., & Gernsheim, A. (1955). The History of Photography.
3. Hannavy, J. (2007). Encyclopedia of Nineteenth-Century Photography. How Light Froze Time Forever
4. Isaacson, W. (2011). Steve Jobs.
5. Newhall, B. (1982). The History of Photography: From 1839 to the Present.
6. Rosenblum, N. (2007). A World History of Photography. Sasson, S. (1975). The First Digital Camera Prototype.

Chapter Nineteen

1. Cook, D. A. (2004). A History of Narrative Film. W. W. Norton & Company. Cousins, M. (2011). The Story of Film: A Worldwide History. Pavilion Books. IMDb. (2024). Highest-Grossing Films.
2. Isaacson, W. (2011). Inventors. Simon & Schuster.
3. Monaco, J. (2000). How to Read a Film: Movies, Media, and Beyond. Oxford University Press.
4. National Science and Media Museum. (2021). Muybridge and Motion.
5. Robinson, D. (1997). From Peepshow to Palace: The Birth of American Film. Columbia University Press.
6. The Thought of Cinema

Chapter Twenty

1. BBC Science Focus. (2023). How the Theremin Works. BBC Science Focus. How the Theremin Works, 2023.
2. Harvard Health Publishing. (2022). How the Brain Controls Speech and Sound. Harvard Health Publishing. Brain and Speech: The Science of Sound, 2022.
3. Isaacson, W. (2023). The Innovators. Simon & Schuster.
4. Isaacson, Walter. The Innovators. Simon & Schuster, 2023.
5. Levitin, D. J. (2006). This Is Your Brain on Music: The Science of a Human Obsession. Dutton.

6. Levitin, Daniel J. This Is Your Brain on Music: The Science of a Human Obsession. Dutton, 2006.
7. Music and the Brain Foundation. (2023). Neurological Pathways of Music and Voice.
8. Music and the Brain Foundation. Vocal Tract and Neurological Imitation in Instrument Design, 2023.
9. National Institute on Deafness and Other Communication Disorders. (2023). How the Voice Works.
10. NIDCD. Understanding the Voice. U.S. Department of Health & Human Services, 2023.
11. Sachs, C. (1940). The History of Musical Instruments. W. W. Norton & Company.
12. Sachs, Curt. The History of Musical Instruments. W. W. Norton & Company, 1940.
13. When the Voice Became Wood, Reed, and Brass
14. Chapter Twenty-One
15. Isaacson, W. (2017). Leonardo da Vinci. Simon & Schuster. The Man Who Dared to See What Others Ignored

Part 2

Chapter One

1. Britannica. (2025). Wheel | Invention, History & Uses. Retrieved from
2. New York Post. (2025). Flying Car Soars Over Traffic in Astonishing First-Ever Video: —New Transportation Is Possible. Retrieved from
3. Voices from History. (2024). How the Invention of the Wheel Changed Society: Transformative Impacts. Retrieved from
4. World History Encyclopedia. (2021). Mesopotamian Inventions. Retrieved from

Chapter Two

1. Encyclopedia Britannica. (n.d.). Isaac Newton. Retrieved from https://www.britannica.com/biography/Isaac-Newton
2. Gleick, J. (2003). Isaac Newton. Pantheon Books.
3. NASA Science. (n.d.). Chapter 3: Gravity & Mechanics. Retrieved from https://science.nasa.gov/learn/basics-of-space-flight/chapter3-2/
4. The Apple That Launched Us into the Stars - (Isaac Newton)

Chapter Three

1. Cambridge University Press – Railways and the Industrial Age, Encyclopedia Britannica – James Watt and Steam Power
2. MIT Technology Review – How Trains Transformed the World
3. National Railway Museum, UK – History of Steam Locomotives Smithsonian Institute – The First Passenger Railways Transportation – The Thought That Moved the World

Chapter Four

1. Carlson, W. Bernard. Tesla: Inventor of the Electrical Age. Princeton University Press, 2013.
2. Davy, H. (1802). —The Electric Arc Experiment.
3. Edison, T. (1879). Development of the Incandescent Lamp. Electricity – The Thought That Conquered Darkness Encyclopaedia Britannica. Michael Faraday Biography.
4. Franklin, B. (1752). Experiments and Observations on Electricity.
5. IEEE Global History Network. History of Electricity. IEEE.org.
6. Jonnes, J. Empires of Light: Edison, Tesla, Westinghouse, and the Race to Electrify the World. Random House, 2003.
7. Morris, E. Edison. Random House, 2019.
8. Tesla, N. (1896). Polyphase System and Niagara Falls Power Plant.
9. Westinghouse, G. (1893). AC Power Transmission and the Chicago World's Fair.

Chapter Five

1. Morris, E. (2019). Edison. Random House.
2. The Relentless Spark That Lit the World - (Thomas Edison)

Chapter Six

1. Cheney, M. (2011). Tesla: Man Out of Time. Simon & Schuster. Nikola Tesla — The Man Who Listened to the Future
2. O'Neill, J. J. (1944). Prodigal Genius: The Life of Nikola Tesla. Ives Washburn. Seifer, M. J. (1996). Wizard: The Life and Times of Nikola Tesla. Citadel Press.

Chapter Seven

1. AT&T Archives. (n.d.). A Century of Communication. Retrieved May 2025.
2. AT&T Archives. (n.d.). The History of the Telephone. Retrieved from https://www.att.com/history
3. Bruce, R. V. (1990). Bell: Alexander Graham Bell and the Conquest of Solitude. Cornell University Press.
4. Bruce, R. V. (1990). Bell: Alexander Graham Bell and the Conquest of Solitude. Ithaca, NY: Cornell University Press.
5. Gray, C. (2006). Alexander Graham Bell and the Conquest of Solitude. Cambridge, MA: Harvard University Press.
6. Gray, C. (2006). Alexander Graham Bell and the Conquest of Solitude. Harvard University Press.
7. Isaacson, W. (2011). The Innovators. New York, NY: Simon & Schuster.
8. Isaacson, W. (2011). The Innovators: How a Group of Hackers, Geniuses, and Geeks Created the Digital Revolution. Simon & Schuster.
9. The Library of Congress. (n.d.). Alexander Graham Bell Family Papers. Retrieved from https://www.loc.gov/collections/alexander-graham-bell- papers

10. The Library of Congress. (n.d.). Alexander Graham Bell Papers. Retrieved May 2025.
11. The Thought of Telephones – Voices Across Wires

Chapter Eight

1. Daimler AG. (n.d.). Bertha Benz: The first long-distance driver in the world. Encyclopedia Britannica. (n.d.). Karl Benz and the invention of the automobile.
2. Lacey, R. (1986). Ford: The Men and the Machine. Mom, G. (2014). Atlantic Automobilism.
3. Tesla, Inc. (n.d.). Impact Report.
4. The Thought of Automobiles – Personal Freedom

Chapter Nine

1. Design Museum. (2015). Fifty Cars That Changed the World. Conran Octopus.
2. Heller, S. & Vienne, V. (2001). 100 Ideas that Changed Graphic Design. Laurence King Publishing.
3. History.com Editors. (2018). Swastika History. History.com. Isaacson, W. (2011). Steve Jobs. Simon & Schuster.
4. Petroski, H. (1992). The Evolution of Useful Things. Vintage. References and Bibliography
5. Symbols: The Silent Messengers of Our Thoughts

Chapter Ten

1. Crouch, T. D. (2002). The Bishop's Boys: A Life of Wilbur and Orville Wright.
2. W. W. Norton & Company.
3. FAA Historical Aviation Milestones
4. Gibbs-Smith, C. H. (2003). The Invention of the Aeroplane 1799–1909. HMSO.
5. McCullough, D. (2015). The Wright Brothers. Simon & Schuster. NASA Aeronautics Timeline
6. Smithsonian National Air and Space Museum Archives
7. They Made a Toy Given by Their Father a Reality for the World

Chapter Eleven

1. Bown, Stephen R. The Last Viking: The Life of Roald Amundsen. Da Capo Press, 2012.
2. Compton-Hall, Richard. The Submarine Pioneers. Sutton Publishing, 1999.
3. International Maritime Organization (IMO): —Ship Design and Construction, 2023.
4. Meyer Werft Shipyard Archives, Germany.
5. Ocean Giants: Building a Ship That Floats, Breaths, and Moves Cities References and Bibliography
6. Stopford, Martin. Maritime Economics. Routledge, 3rd Edition, 2009.

7. Inside the World's Largest Shipyards. Discovery Channel Documentary, 2021.
8. Mega Machines: Sea Giants. National Geographic, 2022.
9. Shipbuilding: How It Works. BBC Future, 2023.

Chapter Twelve

1. BBC. (n.d.). Who invented the traffic light? BBC History. https://www.bbc.co.uk
2. Federal Aviation Administration. (n.d.). A brief history of air traffic control. https://www.faa.gov
3. Hobbs, D. (2018). Aircraft and air traffic control: History and future trends. Aviation Press.
4. International Maritime Organization. (n.d.). History of IMO and maritime safety. https://www.imo.org
5. McShane, C. (1994). Down the asphalt path: The automobile and the American city. Columbia University Press.
6. National Air and Space Museum. (n.d.). Development of air traffic control. Smithsonian Institution. https://airandspace.si.edu
7. National Museum of American History. (n.d.). Transportation collections: Traffic and mobility. Smithsonian Institution. https://americanhistory.si.edu
8. Norton, P. D. (2008). Fighting traffic: The dawn of the motor age in the American city. MIT Press.
9. Traffic Control Systems
10. U.S. Coast Guard Navigation Center. (n.d.). Maritime traffic management history. https://www.navcen.uscg.gov
11. Wright Brothers Aeroplane Company. (n.d.). The 1903 Wright Flyer and the first powered flight. https://www.wright-brothers.org

Chapter Thirteen

1. Department for Transport (UK) – History of Traffic Control Systems, 2015, IEEE Spectrum – Smart Cities and Intelligent Traffic Systems, 2019
2. Los Angeles Department of Transportation – ATSAC System Overview, 2021
3. Norton, Peter. – Fighting Traffic: The Dawn of the Motor Age in the American City, MIT Press, 2008
4. SCOOT Urban Traffic Control – Transport for London Singapore Land Transport Authority – Smart Mobility 2030, 2022
5. The Evolution of Road Traffic Control: From Whistles to Artificial Intelligence

Chapter Fourteen

1. British Railway Museum – The History of Railway Signaling
2. Rail Traffic Management in Europe: A Historical Review‖ – European Railway Agency
3. Railway Gazette – —ERTMS: The Future of Railway Control‖, 2018

4. IEEE Transactions on Intelligent Transportation Systems – —AI in Modern Rail Networks‖, 2021
5. Japanese Ministry of Transport – Bullet Train Safety Systems Overview, 2005 Transport for London – A Century of Signaling the Underground
6. China Railway – Modern High-Speed Train Operations Manual, 2022
7. The Evolution of Rail Traffic Control: From Hand Signals to Autonomous Railways

Chapter Fifteen

1. Paine, Lincoln – The Sea and Civilization: A Maritime History of the World, Alfred A. Knopf, 2013.
2. International Maritime Organization (IMO) – Automatic Identification Systems (AIS) and Maritime Traffic Management Guidelines, www.imo.org
3. United States Coast Guard (USCG) – Marine Navigation and Vessel Traffic Services, Training Manual, 2018.
4. Lighthouse Directory, University of North Carolina – Historical Records of Lighthouses Worldwide, 2023.
5. MarineTraffic – The Rise of Digital Ship Tracking and AIS Technology, 2020 Report.
6. Port of Rotterdam Authority – Smart Port and AI-Driven Vessel Management Systems, Case Study, 2022.
7. National Maritime Museum (UK) – History of Nautical Communication and Flag Signaling, Exhibition Archive, 2021.
8. The Evolution of Sea Traffic Control: From Lighthouses to Global Maritime Intelligence

Chapter Sixteen

1. Eurocontrol – Artificial Intelligence in Air Traffic Management, 2021 Federal Aviation Administration (FAA) – History of Air Traffic Services
2. ICAO (International Civil Aviation Organization) – Evolution of Global Air Traffic Management, 2015
3. MITRE Corporation – ADS-B and Its Role in the NextGen Airspace, 2020
4. NextGen FAA – The Future of Air Traffic: AI and Automation, 2022 Nolan, M. S. – Fundamentals of Air Traffic Control, Cengage Learning, 2010
5. Smithsonian Air and Space Museum – Milestones in Aviation, Navigation, and Control
6. The Evolution of Air Traffic Control: From Ground Flags to AI in the Sky

Chapter Seventeen

1. Chaikin, A. (2007). A Man on the Moon: The Voyages of the Apollo Astronauts. Penguin Books.
2. The Thought of Space Exploration – Beyond the Earth's Boundaries

Chapter Eighteen

1. Isaacson, W. (2014). The Innovators.
2. NASA archives on Sputnik and satellite evolution. Satellites – Reaching for the Stars
3. Siddiqi, A. A. (2000). Challenge to Apollo: The Soviet Union and the Space Race.
4. SpaceX's Starlink deployment reports (2020–2024).

Chapter Nineteen

1. BBC News – —ESA satellite dodges Starlink broadband constellation‖ (2019) Bibliography:
2. European Space Agency (ESA) – Space Debris Office Reports Gorman, Alice. Dr Space Junk vs the Universe. MIT Press, 2019.
3. Johnson, Nicholas L. —Orbital Debris: The Growing Threat to Space Operations. NASA History Series, 2010.
4. LeoLabs – Commercial Space Traffic Data Services
5. Pelton, Joseph N. Space Debris and Other Threats from Outer Space. Springer, 2013.
6. The Need for Traffic Control in Space
7. The Verge – —SpaceX's satellite internet project and the race to dominate low Earth orbit‖ (2020)
8. U.S. Space Command – Space Surveillance Network (SSN)
9. United Nations Office for Outer Space Affairs (UNOOSA) – Space Debris Mitigation Guidelines
10. Weeden, Brian. —Space Traffic Management: Time for Action. Secure World Foundation, 2021.

Chapter Twenty

1. Boston Dynamics. (n.d.).
2. Isaacson, W. (2014). The Innovators. Simon & Schuster.
3. Kaplan, J. (2015). Humans Need Not Apply: A Guide to Wealth and Work in the Age of Artificial Intelligence. Yale University Press.
4. McCarthy, J., Minsky, M., Rochester, N., & Shannon, C. E. (1955). A Proposal for the Dartmouth Summer Research Project on Artificial Intelligence.
5. The Thought of Robots – Machines That Think
6. Turing, A. M. (1950). Computing Machinery and Intelligence. Mind, 59(236), 433–460.

Chapter Twenty-One

1. How a Huge Airplane is Built
2. How Airplanes Are Made‖, Boeing Official Website, www.boeing.com
3. The Birth of the Airbus A380‖, Airbus, www.airbus.com
4. Anderson, John D. Introduction to Flight, McGraw-Hill Education, 2015.

5. How Big Planes Are Builtǁ, Popular Mechanics, 2020.
6. Boeing. (n.d.). Commercial Aircraft Manufacturing. Retrieved from https://www.boeing.com
7. Airbus. (n.d.). A380 Assembly Process. Retrieved from https://www.airbus.com
8. John D. Anderson. Introduction to Flight. McGraw-Hill Education, 2015. Popular Mechanics. (2020). How Big Planes Are Built.

Chapter Twenty-Two

1. Gibson, L. J. (2006). Woodpecker pecking: How woodpeckers avoid brain injury. Journal of Zoology, 270(3), 462–465.
2. Li, Y., & Wang, L. (2015). Bio-inspired design of a shock-absorbing system based on the woodpecker's head. Smart Materials and Structures, 24(9), 095038.
3. Song, Z., Wang, L., & Fan, Y. (2013). Protective design of data recorders inspired by the woodpecker's head. Journal of Bionic Engineering, 10(2), 241– 249.
4. The Bird That Banged Its Head—and Taught Us How to Survive Crashes
5. Wang, L., Cheung, J. T., Pu, F., Li, D., Zhang, M., & Fan, Y. (2011). Why do woodpeckers resist head impact injury: A biomechanical investigation. PLoS ONE, 6(10), e26490.
6. Yoon, S.-H., & Park, S. (2011). A mechanical analysis of woodpecker drumming and its application to shock-absorbing systems. Bioinspiration & Biomimetics, 6(1), 016003.

Chapter Twenty-Three

1. Benyus, J. M. (1997). Biomimicry: Innovation Inspired by Nature. HarperCollins.
2. Garrity, M. (2013). —How a Bird Helped Shape Japan's Bullet Train. Smithsonian Magazine
3. Nakatsu, E. (2009). Biomimicry in Railway Technology. JR West Technical Review.
4. Reynolds, G. (2008). Presentation Zen: Simple Ideas on Presentation Design and Delivery. New Riders.
5. Vincenti, W. G. (1990). What Engineers Know and How They Know It. Johns Hopkins University Press.
6. West Japan Railway Company. Shinkansen Development Notes. BBC News. —The Bird That Inspired Japan's Bullet Train. 2019.

Chapter Twenty-Four

1. Drone Technology – Eyes in the Sky
2. Isaacson, W. (2014). The Innovators. Simon & Schuster.
3. Isaacson, W. (2014). The Innovators: How a Group of Hackers, Geniuses, and Geeks Created the Digital Revolution. Simon & Schuster.
4. McNeal, G. S. (2014). Drones and Aerial Surveillance. Brookings Institution.
5. McNeal, G. S. (2014). Drones and Aerial Surveillance: Considerations for Legislators. Brookings Institution.

6. NASA. (2021). Dragonfly Mission Overview. NASA.gov.
7. NASA. (2021). Dragonfly: A Rotorcraft Lander for Titan Exploration. Retrieved from
8. Singer, P. W. (2009). Wired for War. Penguin Press.
9. Singer, P. W. (2009). Wired for War: The Robotics Revolution and Conflict in the 21st Century. Penguin.

Chapter Twenty-Five

1. Elon Musk — The Mind Rewiring Tomorrow, Isaacson, W. (2023). Elon Musk. Simon & Schuster. Musk, E. (Various Interviews and Speeches).
2. Vance, A. (2015). Elon Musk: Tesla, SpaceX, and the Quest for a Fantastic Future. HarperCollins.

Part 3

Chapter One

1. Britannica. (2025). Wheel | Invention, History & Uses. Retrieved from Computers
2. New York Post. (2025). Flying Car Soars Over Traffic in Astonishing First-Ever Video: —New Transportation Is Possible. Retrieved from
3. Voices from History. (2024). How the Invention of the Wheel Changed Society: Transformative Impacts. Retrieved from
4. World History Encyclopedia. (2021). Mesopotamian Inventions. Retrieved from

Chapter Two

1. Copeland, B. J. (Ed.). (2004). The Essential Turing: Seminal Writings in Computing, Logic, Philosophy, Artificial Intelligence, and Artificial Life. Oxford University Press.
2. Hodges, A. (2014). Alan Turing: The Enigma. Princeton University Press.
3. Leavitt, D. (2006). The Man Who Knew Too Much: Alan Turing and the Invention of the Computer. W. W. Norton & Company.
4. The Man Who Saw Through the Machine
5. Turing, A. M. (1936). On Computable Numbers, with an Application to the Entscheidungsproblem. Proceedings of the London Mathematical Society.

Chapter Three

1. Mead, C. (1990). Neuromorphic electronic systems. Proceedings of the IEEE, 78(10), 1629–1636.
2. The CPU – A Modern Mimic of the Human Brain
3. Von Neumann, J. (1958). The Computer and the Brain. New Haven: Yale University Press.
4. Von Neumann, J. (1958). The Computer and the Brain. Yale University Press.

Chapter Four

1. Essinger, J. (2014). Ada's Algorithm: How Lord Byron's Daughter Ada Lovelace Launched the Digital Age. Melville House.
2. Stein, D. (1985). Ada: A Life and Legacy. MIT Press.
3. The Countess Who Imagined the Future
4. Toole, B. A. (1992). Ada, the Enchantress of Numbers: A Selection from the Letters of Lord Byron's Daughter and Her Description of the First Computer. Strawberry Press.

Chapter Five

1. Berners-Lee, T. (1999). Weaving the Web: The Original Design and Ultimate Destiny of the World Wide Web by Its Inventor. Harper San Francisco.
2. Hafner, K., & Lyon, M. (1996). Where Wizards Stay Up Late: The Origins of the Internet. Simon & Schuster.
3. Isaacson, W. (2014). The Innovators: How a Group of Hackers, Geniuses, and Geeks Created the Digital Revolution. Simon & Schuster.
4. Leiner, B. M., Cerf, V. G., Clark, D. D., Kahn, R. E., Kleinrock, L., Lynch, D. C., ... & Wolff, S. (2009). A brief history of the Internet. ACM SIGCOMM Computer Communication Review, 39(5), 22-31.
5. Markoff, J. (2005). What the Dormouse Said: How the Sixties Counterculture Shaped the Personal Computer Industry. Penguin Books.
6. The Invention of the Internet
7. Tomlinson, R. (2012). The First Network Email. Retrieved from
8. https://www.raytomlinson.com

Chapter Six

1. Holyoak, K. J. (2019). The Human Edge: Analogy and the Roots of Creative Intelligence. Cambridge, MA: MIT Press.
2. Holyoak, K. J. (2019). The Human Edge: Analogy and the Roots of Creative Intelligence. MIT Press.
3. Human Language and Computer Language – A Surprising Resemblance
4. Smith, G. W. (1991). Computers and Human Language. Oxford University Press.
5. Smith, G. W. (1991). Computers and Human Language. Oxford: Oxford University Press.

Chapter Seven

1. Agrawal, A., Gans, J., & Goldfarb, A. (2018). Prediction Machines: The Simple Economics of Artificial Intelligence. Harvard Business Review Press.
2. Cooper, M. (2021). Cutting the Cord: The Cell Phone Has Transformed Humanity. Rosetta Books.

3. Goggin, G. (2006). Cell Phone Culture: Mobile Technology in Everyday Life. Routledge.
4. Heitkoetter, T. (2019). Smartphones: The Real Digital Revolution. Penguin Briefs.
5. Isaacson, W. (2011). Steve Jobs. Simon & Schuster.
6. Ling, R. (2004). The Mobile Connection: The Cell Phone's Impact on Society. Morgan Kaufmann.
7. Mobile Phones – A Whole World in Your Pocket
8. Turkle, S. (2011). Alone Together: Why We Expect More from Technology and Less from Each Other. Basic Books.
9. West, D. M. (2012). The Evolution of Mobile Phones: From the First Call to the Smartphone. Brookings Institution.
10. Zuboff, S. (2019). The Age of Surveillance Capitalism. PublicAffairs.

Chapter Eight

1. Isaacson, W. (2011). Steve Jobs. Simon & Schuster. Kahney, L. (2008). Inside Steve's Brain. Portfolio.
2. Schlender, B., & Tetzeli, R. (2015). Becoming Steve Jobs. Crown Business. Steve Jobs — The Visionary Who Saw Tomorrow Today

Chapter Nine

1. Agar, J. (2013). Constant Touch: A Global History of the Mobile Phone. Icon Books.
2. Bellis, M. (2020). —The History of the Telegraph. ThoughtCo. Retrieved from Isaacson, W. (2011). Steve Jobs. Simon & Schuster.
3. McGregor, C. (2004). Wi-Fi: A New Age of Connectivity. IEEE Communications Magazine.
4. Standage, T. (1998). The Victorian Internet: The Remarkable Story of the Telegraph and the Nineteenth Century's Online Pioneers. Walker & Co.
5. When a Rock Pulled a Nail

Chapter Ten

1. Carlson, W. Bernard. Technology in World History: Volume 5: The Emergence of Modern Communications. Oxford University Press, 2005.
2. Guglielmo Marconi – The Man Who Whispered Through the Air
3. Hong, Sungook. Wireless: From Marconi's Black-Box to the Audion. MIT Press, 2001.
4. J. J. Fahie. Marconi: The Man and His Wireless. Dodd, Mead & Co., 1902. Marconi Company Archives, Bodleian Library, Oxford University. NobelPrize.org–Nobel Prize in Physics 1909:
5. https://www.nobelprize.org/prizes/physics/1909/marconi/biographical/

Chapter Eleven

1. Dean, A. (2017). Hidden Figures of Science: Hedy Lamarr, Inventor and Hollywood Icon. Yale Scientific Magazine.
2. Nash, K. (2020). Frequency Hopping: The Secret Weapon of Hedy Lamarr and George Antheil. Smithsonian Magazine.
3. Rhodes, R. (2011). Hedy's Folly: The Life and Breakthrough Inventions of Hedy Lamarr, the Most Beautiful Woman in the World. Doubleday.
4. The Star Who Invented the Signal of the Future

Chapter Twelve

1. Defence Advanced Research Projects Agency (DARPA). (2023). Navigation without satellites: The future of positioning systems.
2. Einstein, A. (1916). The foundation of the general theory of relativity. Annalen der Physik, 49(7), 769–822.
3. GPS.gov. (n.d.). GPS applications and history. https://www.gps.gov/applications/
4. Kaplan, E. D., & Hegarty, C. J. (2017). Understanding GPS/GNSS: Principles and applications. Artech House.
5. Misra, P., & Enge, P. (2006). Global positioning system: Signals, measurements, and performance. Ganga-Jamuna Press.
6. NASA. (2022). How does GPS work? https://www.nasa.gov/directorates/heo/gps/
7. National Aeronautics and Space Administration (NASA). (2022). The role of GPS in space missions. https://www.nasa.gov/gps-space
8. National Geospatial-Intelligence Agency. (2023). Future of GPS and location-based technologies. https://www.nga.mil/gps-future
9. Parkinson, B. W., & Spilker, J. J. (1996). Global positioning system: Theory and applications (Vols. 1 & 2). American Institute of Aeronautics and Astronautics.
10. Seeber, G. (2003). Satellite geodesy: Foundations, methods, and applications. Walter de Gruyter.
11. Smithsonian Institution. (2022). History of GPS: From Sputnik to smartphones. https://www.si.edu/gps-history
12. Soviet Space Program. (1957). Sputnik 1 launch and tracking experiments. Moscow Space Archives.
13. Thought That Changed How We Navigate the World
14. U.S. Air Force. (2023). The role of GPS in military operations. https://www.airforce.mil/gps
15. U.S. Government. (2023). How GPS works. https://www.gps.gov/howgpsworks/

Chapter Thirteen

1. Berners-Lee, T. (1999). Weaving the Web: The Original Design and Ultimate Destiny of the World Wide Web by Its Inventor. Harper San Francisco.

2. Father Gave Him a Book, and He Gave a Book to the World Internet Live Stats. (2025). World Wide Web Size. Retrieved from
3. Isaacson, W. (2014). The Innovators: How a Group of Hackers, Geniuses, and Geeks Created the Digital Revolution. Simon & Schuster.
4. Pew Research Center. (2023). Internet/Broadband Fact Sheet. Retrieved from References and Bibliography

Chapter Fourteen

1. Levy, S. (2011). In the Plex: How Google Thinks, Works, and Shapes Our Lives. Simon & Schuster.
2. References:
3. Isaacson, W. (2014). The Innovators: How a Group of Hackers, Geniuses, and Geeks Created the Digital Revolution. Simon & Schuster.
4. The Dream That Rewired the World
5. Vise, D. A., & Malseed, M. (2005). The Google Story. Delta Trade Paperbacks.

Chapter Fifteen

1. Apple Inc. (2007, January 9). Apple reinvents the phone with iPhone [Press release].
2. Hurst, S. (1975). The development of the resistive touchscreen. Elographics, Inc.
3. Jobs, S. (2007, January 9). iPhone launch keynote. Macworld Conference. Apple Inc.
4. Johnson, E. A. (1967). Touchscreen technology for air traffic control (Royal Radar Establishment Technical Report). UK: Royal Radar Establishment.
5. Lee, M. (2013). The iPhone and the future of multi-touch technology. Technology Review.
6. Stumpe, B., & Beck, F. (1973). Capacitive touch screen development at CERN. CERN Archives.
7. The Thought to Touch What You See

Chapter Sixteen

1. Bellis, M. (2020). The History of the Telephone. ThoughtCo.
2. Gold, R. (2014). The Race for the Voice: A Brief History of Speech Recognition. IEEE Spectrum
3. Isaacs, K. (2023). How VoIP Works. TechTarget Network. Isaacson, W. (2011). Steve Jobs. Simon & Schuster.
4. Oliphant, R. (2016). How Voice Travels Over Cell Phones. Science ABC.
5. Rabiner, L., & Juang, B. (1993). Fundamentals of Speech Recognition. Prentice Hall.
6. Turing, A. (1950). Computing Machinery and Intelligence. Mind, 59(236), 433– 460.
7. When I Speak into My Phone: A Thought That Travels

Chapter Seventeen

1. Alter, A. (2017). Irresistible: The Rise of Addictive Technology and the Business of Keeping Us Hooked. Penguin Press.
2. Boyd, D. (2014). It's Complicated: The Social Lives of Networked Teens. Yale University Press.
3. Isaac, M. (2019). Super Pumped: The Battle for Uber. W. W. Norton & Company.
4. Kirkpatrick, D. (2010). The Facebook Effect: The Inside Story of the Company That Is Connecting the World. Simon & Schuster.
5. Lanier, J. (2018). Ten Arguments for Deleting Your Social Media Accounts Right Now. Henry Holt and Co.
6. Marwick, A. E. (2013). Status Update: Celebrity, Publicity, and Branding in the Social Media Age. Yale University Press.
7. Social Media – Connecting Every Human Being Instantly
8. Taplin, J. (2017). Move Fast and Break Things: How Facebook, Google, and Amazon Cornered Culture and Undermined Democracy. Little, Brown.
9. Vaidhyanathan, S. (2018). Antisocial Media: How Facebook Disconnects Us and Undermines Democracy. Oxford University Press.
10. Zuboff, S. (2019). The Age of Surveillance Capitalism. PublicAffairs.

Chapter Eighteen

1. Berners-Lee, T., & Fischetti, M. (1999). Weaving the Web: The Original Design and Ultimate Destiny of the World Wide Web by Its Inventor. Harper San Francisco.
2. Ceruzzi, P. E. (2012). Computing: A Concise History. MIT Press.
3. Dyson, G. (2012). Turing's Cathedral: The Origins of the Digital Universe. Vintage.
4. Isaacson, W. (2014). The Innovators: How a Group of Hackers, Geniuses, and Geeks Created the Digital Revolution. Simon & Schuster.
5. Lanier, J. (2010). You Are Not a Gadget: A Manifesto. Knopf. The Thought That Went Digital

Chapter Nineteen

1. Bauman, M. (2021). The History of Speech Recognition. TechSpot.
2. Cherry, C. (1953). Some Experiments on the Recognition of Speech. Journal of the Acoustical Society of America.
3. DeepMind. (2016). WaveNet: A Generative Model for Raw Audio. Isaacson, W. (2014). The Innovators. Simon & Schuster.
4. Lee, K. F. (2018). AI Superpowers: China, Silicon Valley, and the New World Order. Houghton Mifflin Harcourt.
5. Marcus, G. & Davis, E. (2019). Rebooting AI: Building Artificial Intelligence, We Can Trust. Pantheon.

6. Moore, R. K. (2017). Is Talking to Machines Like Talking to Humans? Computer Speech & Language.
7. Saygin, A. P., et al. (2000). The Symbol Grounding Problem in AI. Trends in Cognitive Sciences.
8. Sundar, S. S. (2020). Voice in Everyday Life: Implications of AI-Based Assistants. Journal of Human-Computer Interaction.
9. When Machines Found Their Voice: A Whisper, A Wave, A Wonder

Chapter Twenty

1. Antonopoulos, A. M. (2017). Mastering Bitcoin: Programming the Open Blockchain (2nd ed.). O'Reilly Media.
2. Blockchain – A Ledger of Thought
3. Casey, M. J., & Vigna, P. (2018). The Truth Machine: The Blockchain and the Future of Everything. New York: St. Martin's Press.
4. Isaacson, W. (2011). Steve Jobs. Simon & Schuster.
5. Nakamoto, S. (2008). Bitcoin: A Peer-to-Peer Electronic Cash System. Retrieved from https://bitcoin.org/bitcoin.pdf
6. Rosic, A. (n.d.). What is Blockchain Technology? [Explainer]. Retrieved from https://blockgeeks.com/guides/what-is-blockchain-technology/
7. Tapscott, D., & Tapscott, A. (2016). Blockchain Revolution: How the Technology Behind Bitcoin and Other Cryptocurrencies Is Changing the World. Portfolio/Penguin.
8. World Economic Forum. (2020). Global Future Council on Cryptocurrencies Report. Retrieved from

Chapter Twenty-One

1. 3D Printing – Creating Reality Layer by Layer
2. Berman, B. (2012). 3-D printing: The new industrial revolution. Business Horizons, 55(2), 155–162.
3. Gibson, I. et al. (2015). Additive Manufacturing Technologies (2nd ed.). Springer.
4. Hull, C. (1986). US Patent No. 4,575,330: Apparatus for production of three-dimensional objects by stereolithography.
5. Kodama, H. (1981). Three-dimensional modelling using photo-hardening polymer. Rev. of Scientific Instruments, 52(11), 1770–1773.
6. Lipson, H., & Kurman, M. (2013). Fabricated: The New World of 3D Printing. John Wiley & Sons.
7. Mironov, V. et al. (2009). Organ printing. Trends in Biotechnology, 27(9), 505– 512.
8. NASA (2019). 3D Printing in Space. Retrieved from

Chapter Twenty-Two

1. Ashton, K. (2009). That Internet of Things' Thing. RFID Journal.
2. Evans, D. (2011). The Internet of Things: How the Next Evolution of the Internet is Changing Everything. Cisco.
3. Evans, D. (2012). The Internet of Things: How the Next Evolution of the Internet Is Changing Everything. Cisco Internet Business Solutions Group.
4. Hancke, G. P., & Kuhn, M. G. (2009). An RFID Distance Bounding Protocol. Proceedings of the First ACM Conference on Wireless Network Security.
5. Kranz, M., Holleis, P., & Schreiner, M. (2010). Embedded Interaction: Interacting with the Internet of Things. IEEE Internet Computing, 14(2), 46– 53.
6. Rose, K., Eldridge, S., & Chapin, L. (2015). The Internet of Things: An Overview. Internet Society.
7. The Thought of Internet of Things (IoT) – When Things Started Talking Want, R. (2011). Enabling the Internet of Things. Computer, 44(1), 28–35.

Chapter Twenty-Three

1. Bengio, Y., LeCun, Y., & Hinton, G. (2015). Deep learning. Nature, 521(7553), 436–444.
2. Deep Learning – Going Beyond Human Thinking
3. Goodfellow, I., Bengio, Y., & Courville, A. (2016). Deep Learning. MIT Press.
4. Jumper, J. et al. (2021). Highly accurate protein structure prediction with AlphaFold. Nature, 596(7873), 583–589.
5. Marcus, G., & Davis, E. (2019). Rebooting AI: Building Artificial Intelligence We Can Trust. Pantheon.
6. Mitchell, M. (2019). Artificial Intelligence: A Guide for Thinking Humans. Penguin.
7. Russell, S., & Norvig, P. (2021). Artificial Intelligence: A Modern Approach (4th ed.). Pearson.
8. Schmidhuber, J. (2015). Deep Learning in neural networks: An overview. Neural Networks, 61, 85–117.
9. Silver, D. et al. (2016). Mastering the game of Go with deep neural networks and tree search. Nature, 529(7587), 484–489.

Chapter Twenty-Four

1. Arute, F., et al. (2019). Quantum supremacy using a programmable superconducting processor, Nature.
2. Deutsch, D. (1985). Quantum theory, the Church–Turing principle, and the universal quantum computer. Proceedings of the Royal Society A.
3. Feynman, R. P. (1982). Simulating physics with computers. International Journal of Theoretical Physics.

4. Google AI Quantum. (2019). Quantum supremacy using a programmable superconducting processor. Nature.
5. Grover, L. K. (1996). A fast quantum mechanical algorithm for database search.
6. Isaacson, W. (2021). The Code Breaker. Kaku, M. (2011). Physics of the Future.
7. Preskill, J. (2018). Quantum Computing in the NISQ era and beyond. Quantum Computing – The Next Frontier of Technology Shor, P. W. (1994). Algorithms for quantum computation: Discrete logarithms and factoring.
8. Susskind, L. (2014). Quantum Mechanics: The Theoretical Minimum.

Chapter Twenty-Five

1. Brown, T. et al. (2020). Language Models are Few-Shot Learners. NeurIPS, 33, 1877–1901.
2. ChatGPT – Talking to Machines
3. Marcus, G., & Davis, E. (2019). Rebooting AI: Building Artificial Intelligence We Can Trust. Pantheon.
4. Mitchell, M. (2019). Artificial Intelligence: A Guide for Thinking Humans. Penguin.
5. Ouyang, L., Wu, J., Jiang, X., et al. (2022). Training language models to follow instructions with human feedback. OpenAI Research.
6. Radford, A., Wu, J., Child, R., et al. (2019). Language Models are Unsupervised Multitask Learners. OpenAI Technical Report.
7. Russell, S., & Norvig, P. (2021). Artificial Intelligence: A Modern Approach (4th ed.). Pearson.
8. Smith, B. (2021). Tools and Weapons: The Promise and the Peril of the Digital Age. Penguin Press.
9. Vaswani, A. et al. (2017). Attention Is All You Need. Advances in Neural Information Processing Systems (NeurIPS).

Part 4

Chapter One

1. Exploring the Human Body: A Journey Through the Centuries
2. Harvey, W. (1628). Exercitatio Anatomica de Motu Cordis et Sanguinis in Animalibus. Frankfurt: William Fitzer.
3. Jenner, E. (1798). An Inquiry into the Causes and Effects of the Variolae Vaccinae. London: Sampson Low.
4. Leeuwenhoek, A. (1677). —Observations, communicated to the Royal Society. Philosophical Transactions of the Royal Society.
5. Lister, J. (1867). —On the antiseptic principle in the practice of surgery. The Lancet.
6. Pasteur, L. (1881). —On the germ theory and its applications to medicine and surgery. Bulletin de l'Académie de Médecine.
7. Vesalius, A. (1543). De Humani Corporis Fabrica. Basel: Johannes Oporinus.

8. Watson, J. D., & Crick, F. H. C. (1953). —Molecular structure of nucleic acids: A structure for deoxyribose nucleic acid. Nature, 171(4356), 737–738.

Chapter Two

1. Bellis, M. (2021). The Invention of the Telephone. ThoughtCo.
2. https://www.thoughtco.com/invention-of-the-telephone-alexander-graham- bell-1991383
3. Charles Darwin – The Man Who Let Nature Whisper
4. Goodfellow, I., Bengio, Y., & Courville, A. (2016). Deep Learning. MIT Press.
5. Isaacson, W. (2014). The Innovators: How a Group of Hackers, Geniuses, and Geeks Created the Digital Revolution. Simon & Schuster.
6. LeCun, Y., Bengio, Y., & Hinton, G. (2015). Deep Learning. Nature, 521(7553), 436–444.
7. Marr, B. (2018). Artificial Intelligence in Practice. Wiley.
8. MIT Technology Review. (2020). How Voice Assistants Work.
9. https://www.technologyreview.com

Chapter Three

1. Cobb, M. (2002). Excerpt from The Making of Modern Medicine: Turning Points in the Treatment of Disease.
2. Doudna, J. A., & Sternberg, S. H. (2017). A Crack in Creation: Gene Editing and the Unthinkable Power to Control Evolution.
3. Eknoyan, G. (2008). The legacy of ancient Egyptian medicine. Kidney International, 74(3), 261–264.
4. Magner, L. N. (2005). A History of Medicine. CRC Press. Mukherjee, S. (2016). The Gene: An Intimate History. Scribner. The Thought of Overcoming Limits
5. Topol, E. (2019). Deep Medicine: How Artificial Intelligence Can Make Healthcare Human Again. Basic Books.

Chapter Four

1. Carter, K. C. (2003). The Germ Theory, Medical Innovations, and the History of Handwashing. Cambridge University Press.
2. Fleming, A. (1929). On the Antibacterial Action of Cultures of a Penicillium, with Special Reference to Their Use in the Isolation of B. influenzae. British Journal of Experimental Pathology, 10(3), 226-236.
3. Invention of the Vaccine
4. Nuland, S. B. (2003). Doctors: The Biography of Medicine. Alfred A. Knopf.
5. Pasteur, L. (1881). On the Germ Theory and Its Applications to Medicine and Surgery. Bulletin de l'Académie de Médecine, 7(1), 360-365.

6. Rosen, G. (2015). A History of Public Health. Johns Hopkins University Press.
7. Semmelweis, I. (1861). Die Ätiologie, der Begriff und die Prophylaxis des Kindbettfiebers. Verlag von C.A. Hartleben.
8. Snow, S. (2017). The Discovery of Penicillin and the Dawn of the Antibiotic Age. Oxford University Press.
9. Stern, A. M., & Markel, H. (2005). The History of Vaccination and Its Critics. The New England Journal of Medicine, 352(23), 2324-2332.

Chapter Five

1. Davenport, H. W. (1990). The History of Anaesthesia. Fenster, J. (2001). Ether Day. HarperCollins.
2. Gawande, A. (2002). Complications: A Surgeon's Notes on an Imperfect Science. Picador.
3. Pain-Free Surgery: The Birth of Anaesthesia

Chapter Six

1. Assmus, A. (1995). Early History of X-rays. Physics Today.
2. Bloch, F. & Purcell, E. (1946). Nuclear Induction. Physical Review.
3. Damadian, R. (1971). Tumor Detection by NMR. Science. Damadian, R. (2001). The Story of MRI. Academic Press.
4. Kevles, B.H. (1997). Naked to the Bone. Rutgers University Press.
5. Lauterbur, P.C. (1973). Image Formation by Induced Local Interactions. Nature.
6. Roentgen, W.C. (1896). On a New Kind of Rays. Nature. Seeing the Invisible: The X-Ray and MRI Revolution

Chapter Seven

1. An Inquiry into Vaccination – Edward Jenner Charaka Samhita – Ayurveda (India)
2. Huangdi Neijing – Traditional Chinese Medicine (China) On the Motion of the Heart – William Harvey
3. The Antiseptic Principle – Joseph Lister
4. The Canon of Medicine – Avicenna (Unani / Islamic Medicine) The Discovery of Penicillin – Alexander Fleming
5. The Germ Theory of Disease – Louis Pasteur & Robert Koch The Rise of Allopathic Medicine

Chapter Eight

1. Blueprints of Ourselves: How Humans Recreated Their Own Bodies Through Machines
2. Hampson, R. E. et al. (2018). Developing a hippocampal neural prosthetic to facilitate human memory encoding and recall. Journal of Neural Engineering, 15(3), 036014.
3. Hampson, R. E. et al. (2018). Journal of Neural Engineering, 15(3), 036014.
4. Herr, H. (2014). The new biomechanics of smart prosthetics. Scientific American, 310(5), 72-77.

5. Isaacson, W. (2011). Steve Jobs. New York: Simon & Schuster. Isaacson, W. (2011). Steve Jobs. Simon & Schuster.
6. National Kidney Foundation. (2024). Dialysis Information.
7. National Kidney Foundation. (2024). Dialysis Information. Retrieved from
8. https://www.kidney.org
9. NIH. (2023). Artificial Heart Research.
10. NIH. (2023). Artificial Heart Research. Retrieved from https://www.nhlbi.nih.gov
11. Shepherd, R. F. et al. (2011). Multigait soft robot. PNAS, 108(51), 20400- 20403.
12. Shepherd, R. F. et al. (2011). Multigait soft robot. Proceedings of the National Academy of Sciences, 108(51), 20400-20403.

Chapter Nine

1. Computers and AI in the Medical Field – Thought Meets Healing
2. Dilsizian, S. E., & Siegel, E. L. (2017). Artificial Intelligence in Medicine: Current Trends and Future Directions. Journal of Nuclear Medicine, 58(1), 1– 5.
3. Gao, J., & Wang, L. (2020). Applications of Artificial Intelligence in Health Care. Healthcare, 8(4), 307.
4. IBM Watson Health. (2020). How IBM Watson is transforming healthcare. Retrieved from
5. Shortliffe, E. H., & Cimino, J. J. (2006). Biomedical Informatics: Computer Applications in Health Care and Biomedicine (3rd ed.). Springer.
6. Topol, E. J. (2019). Deep Medicine: How Artificial Intelligence Can Make Healthcare Human Again. Basic Books.

Chapter Ten

1. Murphy, S.V., & Atala, A. (2014). 3D bioprinting of tissues and organs. Nature Biotechnology, 32(8), 773–785.
2. Noor, N. et al. (2019). 3D Printing of Personalized Thick and Perfusable Cardiac Patches and Hearts. Advanced Science, 6(11), 1900344.
3. Ozbolat, I.T. (2015). Bioprinting scale-up tissue and organ constructs for transplantation. Trends in Biotechnology, 33(7), 395–400
4. Kang, H.W. et al. (2016). A 3D bioprinting system to produce human-scale tissue constructs with structural integrity. Nature Biotechnology, 34, 312–319.
5. Printing the Unprintable — Human Parts from a Machine

Part 5

Chapter One

1. Bernstein, W. J. (2008). A Splendid Exchange: How Trade Shaped the World. Grove Press.
2. Frankopan, P. (2015). The Silk Roads: A New History of the World. Bloomsbury.
3. Isaacson, W. (2011). Steve Jobs. Simon & Schuster.

4. Pomeranz, K., & Topik, S. (2013). The World That Trade Created: Society, Culture, and the World Economy, 1400 to the Present. M.E. Sharpe.
5. Smith, A. (1776). The Wealth of Nations.
6. The Thought of Trade – How We Began to Exchange the World Weatherford, J. (1997). The History of Money. Crown Business.

Chapter Two

1. Davies, G. (2016). A History of Money: From Ancient Times to the Present Day. University of Wales Press.
2. Ferguson, N. (2008). The Ascent of Money: A Financial History of the World. Penguin.
3. Money: From Barter to Blockchain
4. Nakamoto, S. (2009). Bitcoin: A Peer-to-Peer Electronic Cash System. Bitcoin.org.
5. References and Bibliography
6. Tang Dynasty Archives (2023). The First Paper Money and Its Economic Impact. Retrieved from.
7. Weatherford, J. (1997). The History of Money. Crown Business.
8. World Bank Reports (2023). The Future of Digital Currency. Retrieved from.

Chapter Three

1. History of Lydian coins and early currency systems, 7th century BCE
2. Mesopotamian clay tablet debt records and origins of credit. Development of banking networks in medieval Florence, including Medici banking influence on Dutch East India Company shares and Amsterdam stock exchange origins, 1602
3. Establishment and stabilizing role of the Bank of England starting in 1694
4. U.S. Federal Reserve, IMF, and World Bank historical roles. Modern technological transformation: digital payments, blockchain, AI in finance

Chapter Four

1. Ferguson, N. The Ascent of Money (2008), Goetzmann, W. N. Money Changes Everything (2017)
2. Homer, S., & Sylla, R. A History of Interest Rates (2011). Kindleberger, C. P. A Financial History of Western Europe (1993)
3. Rothbard, M. N. A History of Money and Banking in the United States (2002)
4. Temin, P. —Financial Intermediation in the Early Roman Empire, Journal of Economic History (2004)
5. The Evolution of Banking: From a Thought to Civilization's Pillar

Chapter Five

1. Ferguson, N. (2008). The Ascent of Money: A Financial History of the World. Penguin Press.
2. Goetzmann, W. N. (2016). Money Changes Everything: How Finance Made Civilization Possible. Princeton University Press.

3. Kindleberger, C. P. (2005). Manias, Panics, and Crashes: A History of Financial Crises. Wiley.
4. Michie, R. C. (1999). The London Stock Exchange: A History. Oxford University Press.
5. New York Stock Exchange. (n.d.). History of the NYSE. Retrieved from
6. Poitras, G. (2000). The Early History of Financial Economics, 1478-1776. Edward Elgar Publishing.
7. The Evolution of the Share Market System

Chapter Six

1. Kaufman, H. (1996). The History of Wall Street in the Civil War Era. University of Chicago Press
2. Steil, B. (2013). The Battle of Bretton Woods. Princeton University Press, The History of the New York Stock Exchange. New York Stock Exchange Wall Street History. Investopedia
3. Wall Street's Rise from Wooden Wall to Financial Giant Zuckerman, M. (2009). The Greatest Trade Ever. Portfolio

Chapter Seven

1. Ali, M. M., & Moon, K. S. (2007). Structural Developments in Tall Buildings: Current Trends and Future Prospects. Architectural Science Review.
2. Burj Khalifa Official Website. (2023). History and Construction. Retrieved from.
3. Condit, C. W. (1998). The Chicago School of Architecture: A History of Commercial and Public Building in the Chicago Area, 1875–1925. University of Chicago Press.
4. Emporis Database. (2023). World's Tallest Buildings and Their Architectural Impact. Retrieved from.
5. Humanity's Ascent from Shelter to Skyline
6. Otis, E. (1853). On the Safety Elevator and Its Applications. New York Historical Archives.
7. Willis, C. (2007). Form Follows Finance: Skyscrapers and Skylines in New York and Chicago. Princeton Architectural Press.

Chapter Eight

1. Dohrn-Van Rossum, G. (1996). History of the Hour: Clocks and Modern Temporal Orders. University of Chicago Press.
2. Landes, D. S. (1983). Revolution in Time: Clocks and the Making of the Modern World. Harvard University Press.
3. Seiko Watch Corporation. (2023). The History of Quartz Watches. Retrieved from Sobel, D. (1995). Longitude: The True Story of a Lone Genius Who Solved the Greatest Scientific Problem of His Time. Walker & Company.
4. The Thought of Convenience: Measuring Time

Chapter Nine

1. Davidson, C. (1998). The History of Glassmaking: A Chronicle of Glass in Western Europe. London: HarperCollins.
2. Duckworth, E. (2003). The Art of Glassblowing: From Antiquity to Modern Day. Chicago: University of Chicago Press.
3. Glass – The Thought That Captured Light
4. Hume, R. (2002). Glass Through the Ages: From Ancient Egypt to the Modern World. New York: Metropolitan Museum of Art.

Chapter Ten

1. General Electric. (2023). The evolution of home refrigeration. GE Appliances.
2. Kha, Y. C. (2007). Refrigeration: A History of Cold Science. Cambridge University Press.
3. Montreal Protocol. (1987). United Nations Environment Programme.
4. Rees, J. (2013). Refrigeration Nation: A History of Ice, Appliances, and Enterprise in America. Johns Hopkins University Press.

Chapter Eleven

1. Ferrara, M. (2012). The Art of Packaging. Promopress.
2. Freedonia Group. (2020). World Packaging Demand Market Report. Harper, C. A. (2000). Handbook of Plastic Processes. Wiley.
3. Ritchie, H., & Roser, M. (2018). Plastic Pollution. Our World in Data. Robertson, G. L. (2016). Food Packaging: Principles and Practice. CRC Press. Wrapped in Thought – A Quiet Journey Through Packaging
4. Yam, K. L. (2009). The Wiley Encyclopedia of Packaging Technology. Wiley.

Chapter Twelve

1. Kummer, C. (1999). A History of Shaving and the Gillette Razor. The Atlantic.
2. Levitt, S. D., & Dubner, S. J. (2005). Freakonomics: A Rogue Economist Explores the Hidden Side of Everything. William Morrow.
3. Petroski, H. (1992). The Evolution of Useful Things. Alfred A. Knopf.
4. Smith, A. (2008). King C. Gillette: The Man and His Shaving Empire. Beard Publishing.
5. The Blade That Rewrote the Morning

Chapter Thirteen

1. Isaacson, W. (2011). Steve Jobs. Simon & Schuster.
2. Library of Congress. (n.d.). Thomas Edison's Phonograph. Retrieved from loc.gov Morton, D. (2000). Off the Record: The Technology and Culture of Sound Recording in America. Rutgers University Press.
3. Sony Corporation. (2009). Sony Design History: Walkman & Personal Audio. When Sound Learned to Stay Wikimedia Commons. (n.d.). Public domain images of the Sony Walkman, iPod, and phonograph.

Chapter Fourteen

1. Casson, M. (2012). The Rise and Fall of the Roman Empire's Logistic System. Oxford University Press.
2. Chopra, S., & Meindl, P. (2018). Supply Chain Management: Strategy, Planning, and Operation. Pearson.
3. Engels, D. (1980). Alexander the Great and the Logistics of the Macedonian Army. University of California Press.
4. Hesse, M., & Rodrigue, J.-P. (2020). The Geography of Transport Systems. Routledge.
5. Isaacson, W. (2011). Steve Jobs. Simon & Schuster.
6. The Silent Force Behind Civilization – A Thought That Moved the World Waters, D. (2019). Global Logistics and Supply Chain Management. Wiley.

Chapter Fifteen

1. Anderson, K. (2013). Designing for Modern Life: Clarence Saunders and the Piggly Wiggly. University of Tennessee Press.
2. Freidberg, S. (2009). Fresh: A Perishable History. Harvard University Press.
3. Lichtenstein, N. (2009). The Retail Revolution: How Wal-Mart Created a Brave New World of Business. Metropolitan Books.
4. Mayo, J. M. (1993). The American Grocery Store: The Business Evolution of an Architectural Space. Greenwood Press.
5. Strasser, S. (1989). Satisfaction Guaranteed: The Making of the American Mass Market. Pantheon Books.
6. When Shopping Became a Quiet Walk-Through Thought
7. Wrigley, N., & Lowe, M. (2002). Reading Retail: A Geographical Perspective on Retailing and Consumption Spaces. Arnold Publishers.

Chapter Sixteen

1. A Thought in a Cup: How Coffee Stirred the World Awake
2. Pendergrast, Mark. Uncommon Grounds: The History of Coffee and How It Transformed Our World. Basic Books, 2010.
3. Topik, Steven & Clarence-Smith, W.G. The Global Coffee Economy in Africa, Asia, and Latin America, 1500–1989. Cambridge University Press, 2006.
4. Ukers, William H. All About Coffee. The Tea & Coffee Trade Journal Company, 1935.
5. Weinberg, Bennett Alan & Bealer, Bonnie K. The World of Caffeine: The Science and Culture of the World's Most Popular Drug. Routledge, 2001.
6. Wild, Antony. Coffee: A Dark History. W.W. Norton & Company, 2005.

Chapter Seventeen

1. A Leaf That Stirred the World: The Story of Tea
2. Benn, J. A. (2015). Tea in China: A Religious and Cultural History. University of Hawai_i Press.

3. Ellis, M. (2015). Empire of Tea: The Asian Leaf that Conquered the World. Reaktion Books.
4. Heiss, M. L., & Heiss, R. J. (2007). The Story of Tea: A Cultural History and Drinking Guide. Ten Speed Press.
5. Macfarlane, A., & Macfarlane, I. (2004). The Empire of Tea: The Remarkable History of the Plant that Took Over the World. Overlook Press.
6. Smith, A. F. (Ed.). (2007). Oxford Companion to American Food and Drink. Oxford University Press.
7. Twilley, N. (2016). —The Tea Spy. The New Yorker.
8. Ukers, W. H. (1935). All About Tea. Tea and Coffee Trade Journal Company.

Chapter Eighteen

1. Coca-Cola Company. (n.d.). History of the Coca-Cola Santa Claus. Retrieved from.
2. Coca-Cola Company. (n.d.). The Story of the Coca-Cola Santa Claus. Coca-Cola Archives.
3. How One Advertisement Redefined Santa for Generations. Moore, C. C. (1822). A Visit from St. Nicholas.
4. Rogers, N. (1997). Santa Claus: A Biography. New York: Random House. Rogers, N. (1997). Santa Claus: A Biography. Random House.

Chapter Nineteen

1. A Thought That Split the World
2. Rhodes, R. (1986). The Making of the Atomic Bomb. Simon & Schuster.
3. United States Department of Energy. (n.d.). Advantages and Challenges of Nuclear Energy.
4. University of Chicago News. (n.d.). The First Nuclear Reactor Explained.
5. Walker, J. S. (2004). Prompt and Utter Destruction: Truman and the Use of Atomic Bombs against Japan. University of North Carolina Press.
6. Wikipedia contributors. (n.d.). History of nuclear power.

Chapter Twenty

1. The Man Most Guarded – A Glimpse into the World's Highest Protection
2. U.S. Secret Service. (2023). Protective Mission Overview. www.secretservice.gov
3. National Geographic. (2019). Inside the White House: Security Secrets.
4. Bamford, J. (2008). The Shadow Factory: The NSA from 9/11 to the Eavesdropping on America. Anchor Books.
5. PBS Frontline. (2001). Protecting the President. Air Force Public Affairs. (2022). VC-25A Fact Sheet: Air Force One.
6. U,S. Government Accountability Office (GAO). (2022). Federal Protective Services Budget Analysis.
7. Newsweek. (2021). How Air Force One Works: Inside the Flying Oval Office.

Chapter Twenty-One

1. Accidents That Changed the World – When Mistakes Became Masterpieces
2. Burke, J. (1978). Connections. Little, Brown and Company.
3. Fenichel, H. (1996). The Birth of Anaesthesia: Discovery of Ether. Harvard Medical Archives.
4. Gawande, A. (2002). Complications: A Surgeon's Notes on an Imperfect Science. Picador.
5. History of Velcro, Velcro Companies, Official Site. Isaacson, W. (2011). The Innovators. Simon & Schuster.
6. Johnson, S. (2010). Where Good Ideas Come From: The Natural History of Innovation. Riverhead Books.
7. Morton, W. T. G. (1846). Demonstration of Surgical Anesthesia, Massachusetts General Hospital Records.
8. Royston, A. (2004). Everyday Inventions. National Geographic Society. The Discovery of X-rays, NobelPrize.org.
9. Watson, S. (2006). —The Accidental Inventions That Changed the World.‖ Scientific American.

Chapter Twenty-Two

1. Bezos, J. (2010, May 18). Jeff Bezos' 2010 Princeton Baccalaureate speech. Princeton University. https://www.princeton.edu/news/2010/05/30/jeff- bezos-90-princeton-baccalaureate-speech
2. Jeff Bezos – The Relentless Architect of Tomorrow
3. Kantrowitz, A. (2014, November 6). Inside Amazon's bold vision for Alexa. Bloomberg Businessweek. https://www.bloomberg.com/news/articles/2014- 11-06/inside-amazons- bold-vision-for-alexa
4. Kraus, C. (2017). The Everything Store: Jeff Bezos and the age of Amazon (Updated ed.). Little, Brown and Company.
5. Levy, S. (2016, September 16). How Amazon, Google, and Microsoft quietly got behind the voice assistant boom. Wired. https://www.wired.com/2016/09/voice-assistants-alexa- google/
6. Stone, B. (2013). The everything store: Jeff Bezos and the age of Amazon. Hachette Book Group.
7. Wingfield, N. (2017, May 9). Jeff Bezos' vision for Amazon: _The everything store' and beyond.
8. The New York Times. https://www.nytimes.com/2017/05/09/technology/jeff-bezos-amazon. html

Chapter Twenty-Three

1. Kroc, R., & Anderson, R. (1977). Grinding It Out: The Making of McDonald's. New York: St. Martin's Press.
2. Love, J. F. (1995). McDonald's: Behind the Arches. New York: Bantam Books.

3. McDonald's Corporation. (2024). Company History. Retrieved from https://corporate.mcdonalds.com
4. The World's Largest Food Chain – How a Single Stand Became a Global Giant
5. The Final Bridge Bibliography

Chapter Twenty-Four

1. Alter, A. (2017). Irresistible: The Rise of Addictive Technology and the Business of Keeping Us Hooked. New York, NY: Penguin Press.
2. Eyal, N. (2014). Hooked: How to Build Habit-Forming Products. New York, NY: Portfolio/Penguin.
3. Harari, Y. N. (2018). 21 Lessons for the 21st Century. London: Jonathan Cape.
4. Lanier, J. (2018). Ten Arguments for Deleting Your Social Media Accounts Right Now. London: Bodley Head.
5. Turkle, S. (2011). Alone Together: Why We Expect More from Technology and Less from Each Other. New York, NY: Basic Books.
6. Zuboff, S. (2019). The Age of Surveillance Capitalism: The Fight for a Human Future at the New Frontier of Power. New York, NY: PublicAffairs.

Chapter Twenty-Five

1. Burgess, J., & Green, J. YouTube: Online Video and Participatory Culture, 2nd ed. Polity, 2018.
2. Friedman, T. L. The World Is Flat: A Brief History of the Twenty-first Century. Farrar, Straus and Giroux, 2005.
3. Isaacson, W. Steve Jobs. Simon & Schuster, 2011.
4. Kelly, K. The Inevitable: Understanding the 12 Technological Forces That Will Shape Our Future. Viking, 2016.
5. Negroponte, N. Being Digital. Alfred A. Knopf, 1995.
6. Shirky, C. Here Comes Everybody: The Power of Organizing Without Organizations. Penguin Press, 2008.

Chapter Twenty-Six

1. International Federation of Robotics. (2024, November 20). *Global robot density in factories doubled in seven years (World Robotics 2024).* https://ifr.org/
2. Markoff, J. (2012, August 18). *Skilled work, without the worker. The New York Times.* https://www.as.utexas.edu/astronomy/education/fall15/wheeler/secure/Robots_NYT.pdf
3. MxD. (2022, May 26). *What's up with lights-out factories?* https://www.mxdusa.org/
4. Omron & Philips. (2019). *Future-proof assembly concept for high-end shavers* [Customer reference PDF]. https://assets.omron-ap.com/

5. Siemens Digital Industries Software. (2021). *The lights-sparse versus the lights-out factory* [White paper]. https://www.plm.automation.siemens.com/
6. Wikipedia contributors. (2025). *Lights out (manufacturing).* In *Wikipedia.* https://en.wikipedia.org/wiki/Lights_out_(manufacturing)
7. Reuters. (2024, November 20). *China overtakes Germany in industrial use of robots, says report.* https://www.reuters.com/
8. Standard Bots. (2025). *Lights-out manufacturing in 2025: Fully automated factories & dark factories.* https://standardbots.com/blog/lights-out-manufacturing

Epilogue

1. Asimov, I. (1994). Understanding Physics. Barnes & Noble. Bronowski, J. (1973). The Ascent of Man. Little, Brown and Company.
2. Gleick, J. (2011). The Information: A History, a Theory, a Flood. Vintage. Harari, Y. N. (2014). Sapiens: A Brief History of Humankind. Harper.
3. Isaacson, W. (2011). Steve Jobs. Simon & Schuster.
4. Standage, T. (1998). The Victorian Internet. Walker Publishing.
5. Watson, J. D. (1968). The Double Helix: A Personal Account of the Discovery of the Structure of DNA. Scribner.

Image Credits

The following images appear throughout *Waves of Thoughts* – Thoughts That Changed the World. Every effort has been made to accurately acknowledge original creators and source repositories. All images listed below are in the public domain or used with appropriate permissions.

Karl Benz Patent Motorwagen (1886).

Patent drawing "Fahrzeug mit Gasmotorenbetrieb," Deutsches Reichspatent Nr. 37435, issued 2 November 1886. Public domain (patent document). Source: Wikimedia Commons. Appears in Part 2, Chapter 8 (p. 117).

Alan Turing (1951).

Photograph by Elliott & Fry. Public domain. Source: Wikimedia Commons. Appears in Part 3 (Digital Tsunami), Chapter 1 (p. 175).

Ada Lovelace (1838).

Portrait painting by Alfred Edward Chalon. Public domain. Source: Wikimedia Commons. Appears in Part 3 (Digital Tsunami), Chapter 4 (p. 185).

Charles Babbage (c. 1847–1851).

Daguerreotype by Antoine Claudet. Public domain. Source: Wikimedia Commons. Appears in Part 3 (Digital Tsunami), Chapter 4 (p. 186).

Hedy Lamarr (1940).

MGM studio publicity portrait for Comrade X, attributed to Clarence Sinclair Bull. Public domain. Source: Wikimedia Commons. Appears in Part 3 (Digital Tsunami), Chapter 11 (p. 211).

ENIAC (1945).

Photograph of the first electronic general-purpose computer, built at the University of Pennsylvania for the U.S. Army. Public domain. Source: U.S. Army photograph, via Wikimedia Commons. Appears in Part 3 (Digital Tsunami), Chapter 18 (p. 237).

Charles Darwin (c. 1868).

Photograph by Julia Margaret Cameron. Public domain. Source: Wikimedia Commons. Appears in Part 4 (Health, Humanity and Science of Life), Chapter 2 (p. 283).

The Beast – U.S. Presidential State Car.

Photograph of the presidential limousine loaded onto a U.S. Air Force transport aircraft. Public domain. Source: U.S. Air Force / DVIDS. Appears in Part 5 (Thoughts That Built Civilization), Chapter 20 (p. 411).

www.ingramcontent.com/pod-product-compliance
Lightning Source LLC
LaVergne TN
LVHW100247210926
848480LV00003B/846
* 9 7 8 1 7 6 4 2 5 6 1 4 8 *